History of Climate Change

HISTORY OF CLIMATE CHANGE

From the Earth's Origins to the
Anthropocene

Antonello Provenzale

Translated by Alice Kilgarriff

polity

Originally published in Italian as *Coccodrilli al Polo Nord e Ghiacci all'Equatore: Storia del clima della Terra dalle origini ai giorni nostri* © First published in Italy by Rizzoli, 2021
This edition published by arrangement with Grandi & Associati

This English edition © Polity Press, 2023

The translation of this work has been funded by SEPS
SEGRETARIATO EUROPEO PER LE PUBBLICAZIONI SCIENTIFICHE

Via Val d'Aposa 7 – 40123 Bologna – Italy
seps@seps.it – www.seps.it

Polity Press
65 Bridge Street
Cambridge CB2 1UR, UK

Polity Press
111 River Street
Hoboken, NJ 07030, USA

ISBN-13: 978-1-5095-5393-8 – hardback

A catalogue record for this book is available from the British Library.

Library of Congress Control Number: 2022948541

Typeset in in 11.5 on 14 Adobe Garamond
by Fakenham Prepress Solutions, Fakenham, Norfolk NR21 8NL
Printed and bound in Great Britain by CPI Group (UK) Ltd, Croydon

The publisher has used its best endeavours to ensure that the URLs for external websites referred to in this book are correct and active at the time of going to press. However, the publisher has no responsibility for the websites and can make no guarantee that a site will remain live or that the content is or will remain appropriate.

Every effort has been made to trace all copyright holders, but if any have been overlooked the publisher will be pleased to include any necessary credits in any subsequent reprint or edition.

For further information on Polity, visit our website:
politybooks.com

Contents

Introduction

The climate of our planet is continually changing. At certain periods in its history, the Earth was hot and lush with forests, and at others, it was almost entirely covered by a thick layer of ice. At times, there have been crocodiles at the North Pole, and at others, ice at the equator. Over the course of millions of years, sudden rises in temperature caused by massive volcanic eruptions have been followed by slow cooling periods associated with the dance of the continents, moved by convective motions in the Earth's mantle. And all of this has been punctuated, every so often, by monstrous catastrophes that have swept away most of that epoch's living beings.

Over the last 3 million years, the climate has been dominated by the almost regular alternation between prolonged cold and dry glacial periods and shorter, hot and rainy interglacials, in response to slow variations in the Earth's orbit. The time in which we are currently living, the Holocene, began around 12,000 years ago when the ice retreated after the last glacial peak. Unlike many of the periods that preceded it, the Holocene has been characterized up until now by a stable climate with modest variations in temperature, something that has facilitated the birth of agriculture and the shift from nomadic hunter-gatherer populations to sedentary human societies.

The mechanisms that determine the continuous modifications in our climate are multiple and depend on both external forces – such as variations in solar luminosity, Earth's orbital characteristics and volcanic activity – and, most of all, on the many internal processes connecting the atmosphere, the oceans, the ice, the biosphere, the crust and the mantle. These are extremely complex phenomena in which the response is often disproportionate to the stimulus, and it can activate a myriad of feedback processes. They are associated to changes across all scales of time and space, with the amplification or reduction of the effects of external forcing factors. On top of these, the bond between organisms and environment is what makes the Earth a living planet whose evolution is inextricably linked to the action of the biosphere.

Among the many players involved, two play a central role. One is the composition of the atmosphere, in particular the concentration of greenhouse gases such as carbon dioxide, methane and water vapour. The other is the albedo of terrestrial surfaces and the clouds, or, rather, the quantity of solar energy that is reflected into space without entering into the planet's thermal machine. Alongside these principal actors, we also have the transport of heat and matter in the atmosphere and the ocean; the water cycle with its transformation of water into liquid, solid and vapour; and the biogeochemical cycles that remix and redistribute crucial elements such as carbon, phosphorus, nitrogen, iron and many others among the various components of the Earth System.

Over the last two centuries, a new factor has been added to these mechanisms of the climate's natural variability: human activity. This has reached planetary dimensions, emitting enormous quantities of carbon dioxide into the atmosphere and profoundly modifying the land, while eliminating an incredibly large number of natural ecosystems, deforesting the tropics, dramatically reducing terrestrial and marine biodiversity, and moving unimaginable quantities of sediment. All of these changes, in particular the growth in the concentration of carbon dioxide in the atmosphere, have led to a rapid increase in the average global temperature, which has risen by more than 1 degree centigrade over the last century. In some areas, such as the Arctic, the warming has been double that. And these changes in temperature have an impact on all of the climate's components – leading, for example, to an intensification of drought conditions in the Mediterranean basin and higher rainfall in Northern Europe.

Of course, an increase in temperature of almost 1 degree is negligible when compared to the much more conspicuous variations of the past. Some 50 million years ago, the average global temperature was probably 10 degrees higher than today. So, what are we worrying about?

Essentially, two aspects. First, there is the fact that the temperature increase of the last fifty years has been extremely fast, probably faster than has ever occurred in the past, making it more difficult for the environment to adapt to the new conditions. And there is the not-insignificant fact that, as of 2022, almost 8 billion people inhabit the Earth within fixed infrastructures and megalopolises that grow ever larger, often along densely populated coastlines that are the site of intense production activity. This is a population that needs water, food, energy. We are a highly technological society, but one that is very vulnerable to environmental changes. If the mercury continues to climb, there will be more droughts and consequent famines but also, paradoxically, increasingly violent floods, more extensive fires, a rise in sea level and more intense coastal erosion events. Some ecosystems could even collapse, and there will be difficulties in agricultural production. And, presumably, there will also be mass migration, social and economic instability and an even greater risk of war. If, by the end of the century, the temperature has increased by 4 °C compared to 100 years ago, as is expected if we continue to behave in the way we have done so far, the world will not be an easy place to live.

In order to understand fully what is happening, to put it in perspective and attempt to resolve the problems we ourselves have created, we must not deny the factual evidence – global warming is caused by human activity – and nor should we scream that this is the end of the world. Instead, it is necessary to understand how the climate functions, how it has changed and why, what is happening now, what the risks are, and what the past, even the remote past, can teach us. And how we can attempt to predict the future, and what strategies we can use to tackle the climate crisis we are living in. In short, we need to apply our intelligence and capacity for analysis. This book aims to make a small contribution in this regard, discussing what we have understood about the fascinating and complex system that is the Earth's climate, and what remains unclear. It does so by following a path that begins in the darkness of time, in the era when our planet was formed, through conditions so very different from those we see today that they bring to mind alien worlds – all the while, trying to understand why things have happened in a particular way.

We will journey through the history of the climate of the Earth, investigating the mechanisms by which it functions, exploring the most extreme conditions of the past and the great instabilities that have changed the world, finally arriving at the last century, at the cumbersome presence of humanity and the effects caused by the rise in the concentration of carbon dioxide in our atmosphere and oceans. We will do this in order to compare what is happening today with what we have learned from the Earth's past, and to use our knowledge of previous disasters to avoid (where possible) new ones.

The knowledge and capacity to forecast that come from this cannot, however, remain an end in themselves. These things must inform those taking decisions and be used to resolve real problems. At this moment, one of the issues we must face up to is how to combat anthropogenic climate change, as well as the need to stop the loss of biodiversity and fight poverty the world over. These three aspects are strongly connected to one another and involve the health of human beings and ecosystems alike, as well as the availability of water of sufficient quantity and quality for both human beings and the natural environment. In the final chapters of this book, we will consider some of these aspects with the knowledge that the solution is not a return to a golden age that never actually existed, but by taking as a starting point those technologies and that development which created the problem and using them to resolve it. We need to imagine and implement innovative approaches, even more advanced technologies, societies that are less delusional and more sustainable, working with nature and not against it. Preferring utopia to dystopia and trying to create it in reality, at least as far as possible. Not so much to save the planet, which will do just fine on its own and probably better without us, but to save ourselves.

I would like to thank everyone who has helped me to write this book. I am grateful to Giò and Maria for their support during the months of writing. Thank you to Marco Ferrari, Silvia Giamberini, Elisa Palazzi and Maddalena Pennisi, who had the patience to read everything I wrote and have always given me encouragement, advice and relevant suggestions. Thank you to Chiara Boschi, Andrea Dini, Gianfranco Di Vincenzo and Eleonora Regattieri, who provided me with precious guidance on those chapters closest to their own scientific expertise. A special thank you goes to Laura Grandi, who pushed me to take on this task and followed it throughout each of its phases, putting up with my delays and providing me with important advice. Thank you also to the Festival della Mente in Sarzana, Italy, where the idea for this book was born. Finally, sincere thanks to Elise Heslinga of Polity Press, who gently helped me throughout the preparation of the English version of this book and firmly reminded me of deadlines; to Alice Kilgarriff, who translated the book from Italian and with whom I worked closely and pleasantly; and to Leigh Mueller for the copy-editing. Obviously, it goes without saying that all of the errors are, however, exclusively my own.

From the Ocean of Magma to the Great Oxygenation

The Beginnings

Below is an ocean of molten magma the colour of red fire. Above, an incessant rain of meteorites and small asteroids, bolides that crash onto the burning liquid surface of the planet, releasing even more heat and scattering carbon compounds that will later be incorporated into the molecules of life. In the first 200 million years of its existence, the Earth boiled incessantly, resembling the most pulp-like descriptions of a Dantean Hell.

During this first geological eon, appropriately named the Hadean, inside the planet a dense nucleus that was rich in metals was differentiating itself from a lighter, rocky mantle made up of siliceous and partially molten materials. Over this time period that lasted around 500 million years and ended 4 billion years ago, it is believed that the orbit of a planetoid known as Theia, which was almost as large as Mars, most likely crossed into the orbit of our own planet, colliding with the Earth in a crash of colossal proportions. The two celestial bodies fused together and the violence of the impact broke off a gigantic incandescent fragment that solidified over time, forming our now ice-cold satellite: the Moon.

In the millions of years that followed, the surface of the Earth slowly cooled and became solid, covered first in basalt and later in granite, while asteroid impacts decreased, the continents began to form and widescale volcanic eruptions dominated the planetary stage. The Earth's crust was born. The volcanoes freed enormous quantities of thermal energy still present in the bowels of the earth, releasing gas and water that went on to form the primordial atmosphere. Other energy was produced by the decay of radioactive elements that had been incorporated into the mantle during the formation of the Earth. At that time, with the reduction in surface temperature, new oceans of liquid water covered the surface, the atmosphere began its long chemical evolution and the Earth transformed itself into a planet ready to host life. This was the beginning of the Archean Eon, the period that spans from almost 4 to 2.5 billion years ago.

Life on Earth, therefore, appeared relatively early on. Numerous clues suggest the presence of fully developed cellular life more than 3.5 billion years ago – so, shortly after conditions grew less hostile. How life on Earth originated remains one of science's most fascinating enigmas, bound of course to the question of whether living organisms were also able to develop on other planets. But for now, we are interested in what the climate was like in that distant time, how the ocean, atmosphere and planet's surface interacted with one another and what the global climate conditions were like.

The Atmosphere in the Archean

During the Archean, the atmosphere was different from the one we know today. For a long time, scientists believed that the Earth's primitive atmosphere was strongly 'reducing', meaning it was poor in oxygen and rich in molecules that contained hydrogen, such as methane, hydrogen sulphide and ammonia, substances that are, in truth, not very palatable for us humans. In 1952 and 1953, two American researchers, Stanley Miller, at the time a young assistant, and Harold Urey, his professor, carried out an experiment that continues to be hailed as a milestone. Following the intuitions of Alexander Oparin, a Russian biochemist who thirty years earlier had proposed a theory on the biogeochemical origin of life, Miller and Urey inserted a mixture of gas – considered at the time to be similar to the Archean atmosphere – into a sealed container. They added water and caused various sparks inside the container to simulate the effect of lightning.

The result – a surprising one – was that many complex organic molecules were produced, including different kinds of amino acids that are present in living organisms today. The ensuing enthusiasm and emotion were momentous, and it seemed that we were close to under-standing how life on Earth was born.

However, the journey from Miller and Urey's organic compounds to the function of a living cell is a long one. In the decades that followed, other possibilities were explored. For example, the suggestion that life was born around submarine hydrothermal vents, or by the catalytic action of certain kinds of clay on the formation of protocellular membranes. Or that it began in the depths of the Earth's crust, many kilometres below the surface, with organisms able to use methane as an energy source. There are also those who, like Fred Hoyle, a famous British astrophysicist, supported the hypothesis of panspermia, which suggested life was brought to Earth by comets and asteroids hailing from other planets (though this doesn't resolve the problem of how life was formed on those other planets). Today, we know that complex organic molecules can be observed in interstellar clouds and found on those meteorites that fall to Earth. In 2019, for example, a complex sugar, ribose, was identified in two different meteorites. But for now, at least, we are still not able to reconstruct fully the chain of events that led these complex molecules to become the first functioning cell.

Even the Archean atmosphere was not, in reality, as reducing as we had believed up until a few years ago. In 2011, Dustin Trail and two of his colleagues at the Astrobiology centre at the Rensselaer Polytechnic Institute in the state of New York published a scientific article in which they analysed the composition of the oldest minerals present on Earth. The results of the study suggest that the primordial atmosphere was primarily composed of carbon dioxide, sulphur dioxide and water, all of which are very stable molecules, meaning this atmosphere was much less reducing and decidedly different from that hypothesized in the experiment by Miller and Urey. The search for the origins of life on Earth, therefore, continue today and will likely occupy researchers for many years to come.

That distant world was very different from the one we know today. There was no life on the land, only single-celled life (predominantly organisms similar to bacteria and Archaea) in marine waters, and an atmosphere that was different from the one we breathe in today. If one of us were to be transported back to that environment by a time machine, we would suffocate instantly. There was no molecular oxygen in the atmosphere, no ozone layer to protect from the sun's ultraviolet rays. In short, it was an alien planet with an ecosystem based predominantly on anaerobic metabolism, sustained by chemical reactions that can only happen in the absence of oxygen. But this is the environment in which, after a long chain of successive evolutions, life developed, leading to us and the world we know.

Anoxic Oceans

The oceans of the Archean were also different from those we know today. Ocean salinity was probably not all that different, but as there was no free oxygen in the atmosphere, there was none dissolved in the marine waters. The oceans were *anoxic*. Thanks to this total absence of oxygen, the hot waters of the Archean contained high quantities of dissolved soluble iron (known as 'ferrous iron', an iron atom with two positive charges) that came from rocks and which would be immediately oxidized in today's oceans and rendered insoluble, causing them to sink into the sediment. The single-celled organisms present in those seas lived in anaerobic conditions, meaning their metabolism required an almost complete absence of free oxygen and a lifecycle based on chemical reactions between atoms of dissolved iron and carbon dioxide. Other kinds of metabolism used sulphur, which was widely available in the Archean oceans. For example, the process of photosynthesis developed by the so-called purple sulphur bacteria was based on an anaerobic cycle that used sunlight and hydrogen sulphide.

In that world based on the chemistry of iron and sulphur, however, other single-celled organisms were also present, capable of a different kind of photosynthesis that was much more efficient. Cyanobacteria, also known as blue-green algae, were able to use carbon dioxide, water and sunlight from the atmosphere in a new photosynthetic process to generate energy, producing molecular oxygen as a waste material. We know that cyanobacteria were almost certainly already present 2,700 million years ago, perhaps even earlier. In that world, oxygen was a toxic gas, lethal for anaerobic organisms, and these new arrivals would come to expand from the ecological niches in which they had developed and overrun the anoxic world of the Archean.

The presence of dissolved iron and the production of oxygen by cyanobacteria in the oceanic waters of the late Archean Eon gave rise to the spectacular sedimentary rocks known as banded iron formations. These highly colourful structures, central to the mineral extraction of iron, are visible today in the oldest rocks of Australia, Brazil, Canada, South Africa, India, Russia and the United States. The thickness of each band varies from less than a millimetre to a metre, and is composed of hematite or magnetite, both iron oxides. Between these bands we find siliceous sedimentary rocks composed of chert and shale.

But how are these bands formed? We can presume that oxygen production by the cyanobacteria played an essential role. We have seen that, during photosynthesis, oxygen is produced as a waste material. The oxygen produced by the cyanobacteria in the shallow waters of the Archean continental platforms was therefore rapidly eliminated because it combined with the dissolved iron, generating insoluble iron oxides that sank to the bottom and created the rich iron deposits we see today in rocks the world over.

Why, though, was the precipitation process of the iron oxides not continuous, creating alternating layers of sediments, some rich in iron and others from siliceous rocks that instead contain barely any? At the moment, there is no definitive answer. One possibility is that the ecosystem of the cyanobacteria underwent a series of oscillations in which their efficiency varied, meaning their ability to produce oxygen through photosynthesis varied also. Equally, there could have been fluctuations in the chemical characteristics of the environment that rendered it less or more suited to the iron's precipitation. Recently, a group of Australian, German and Canadian researchers proposed a different explanation. According to their hypothesis, supported by an analysis of the microscopic structure of the rocks with banded iron formations, the sedimentation of the iron was roughly continuous. Only after deposition, in the process of diagenesis and metamorphism (which modify sediments and sedimentary rocks when the geological processes lead them to be buried deeper down and exposed to temperatures and pressure much higher than those on the surface), is there a 'migration' of the iron oxides that separate from their siliceous sediment to form alternate bands with a high concentration of iron.

Independently of whether banded iron formations originated from oscillations in the production of oxygen or from the processes that took place after sedimentation, these banded structures tell us two important things. The first is that, 2.5 billion years ago, single-celled organisms capable of photosynthesis, and which produced oxygen as a waste material, already existed. The second is that the oxygen was present in relatively low concentrations, otherwise the iron would not have been able to stay dissolved and would instead have oxidized as soon as it came into contact with the sea water, as happens in modern oceans. It is no coincidence that there are no banded iron formations younger than 1,800 million years old, if we exclude a significant episode around 600 million years ago, to which we will return in the next chapter.

The Kingdom of Oxygen and Sulphur

Around 3 billion years ago, molecular oxygen was almost entirely absent from the Earth's atmosphere and the oceans. Today, the atmospheric concentration of oxygen is around 21 per cent and the oceans are oxygenated to great depths. Something must therefore have happened between that distant past and more recent times that caused the quantity of free oxygen in the atmosphere to grow significantly and drastically modify our planet's environment.

In the relatively shallow waters that covered the continental platforms of the Archean seas, cyanobacteria initially produced small quantities of oxygen that combined with the dissolved iron to produce oxides that were quickly deposited among the sediments on the seabed. The more the cyanobacteria population grew, the more oxygen was produced. Not all of the oxygen was reused by the biological cycle (cyanobacteria, like plants, produce oxygen in the reactions of photosynthesis and reabsorb it during nocturnal respiration). In fact, part of the organic material produced through photosynthesis was buried in the sediment, leaving a surplus of oxygen in the water. As a consequence, the available iron was oxidized and oxygen began to accumulate in the marine water, passing from there into the atmosphere. The anaerobic organisms were confined to marginal areas, beneath the mud on the seabed or at great depths, whilst on the surface the world grew increasingly richer in oxygen.

The atmospheric concentration of oxygen has, nevertheless, grown very slowly since around 2.4 billion years ago. For hundreds of millions of years, oxygen remained well below 10–15 per cent of its current concentration. Figure 1.1, from the work of Donald E. Canfield, a biogeochemist working in Denmark, shows an approximate reconstruction of the level of oxygen in the atmosphere.

There is, however, a 'but'. If the cyanobacteria were the cause of oxygenation and they were already present more than 2.7 billion years ago, why didn't the concentration of oxygen in the atmosphere start to

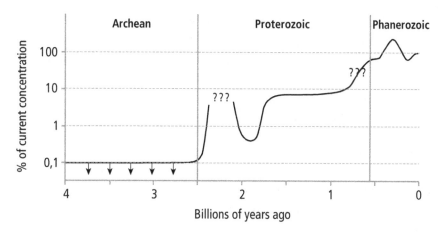

Figure 1.1 Approximate reconstruction of the concentration of oxygen in the atmosphere, as a percentage of the current level, taken from the work of Donald E. Canfield (2015).

increase earlier, as soon as they evolved? Why did it instead remain low for so long?

It is difficult to answer this question with any certainty, because the relevant information we have is still scarce. In any case, there are two possible factors that could have limited the concentration of oxygen: scant production by a still scarcely abundant cyanobacteria population, or a rapid elimination of the oxygen in the atmosphere. Or a combination of both of these things. And, more generally, the chemistry of the ocean and the atmosphere necessarily played an essential role in determining the composition of that world.

One possibility is that in the Archean seas there was little phosphorus, an essential nutrient for cyanobacteria and algae, both then and now. In fact, phosphorus was easily absorbed by the iron oxides that sank, creating the banded iron formations (BIF) we spoke of earlier. The phosphorus dissolved in the marine waters would have therefore been far less than the levels found today, significantly limiting the productivity of the cyanobacteria's ecosystem and, consequently, the production of oxygen through photosynthesis. A cycle of cause and effect would therefore have set in, by which the cyanobacteria produced oxygen that combined with the dissolved iron, causing it to sink to the seabed as iron oxides. The latter, however, absorbed phosphorus, removing it from the water and thus limiting the nutrients and growth of the cyanobacteria and, therefore, the production of oxygen. Such a cycle could have also given rise to temporary oscillations in the entire process, generating the ferrous bands of sediments.

Around 2.4 billion years ago, however, a new event may have interrupted the oxygen–iron–phosphorus cycle. Analysis of sediments and rocks indicates that, at this time, the ocean grew rich in sulphur, which combined with the iron dissolved in the water to form pyrite, with its chemical formula FeS_2 (1 iron atom and 2 sulphur atoms). The pyrite sank to the seabed, removing the iron from the water. At that point, the formation of iron oxides and BIF was significantly reduced, phosphorus was no longer eliminated from the water, nutrients became abundant and the cyanobacteria could expand and dominate the ecosystem, leading to a gradual rise in the concentration of oxygen in the water and the atmosphere.

Magma from the Depths

In order to get a full picture, we must not neglect another fundamental piece of the puzzle: the direct removal of oxygen from the water and the atmosphere.

The oxygen is removed from the Earth's surface in two main ways: through the oxidation of the rocks and organic matter on the planet's surface, and through chemical reactions with reducing gases emitted by volcanoes, which come from deep in the Earth's crust and mantle. Both of these processes were active in the Archean (as they still are today), but at a certain point, when oxygen began to accumulate in the atmosphere, they evidently became less efficient.

In particular, it is probable that the Earth's upper mantle was initially in a chemically reducing state capable of reacting with other compounds and 'capturing' oxygen atoms. The gases produced by volcanic activity, which came from deep magma, were also reducing and rich in hydrogen compounds able to react with and absorb the oxygen from the atmosphere. One possible scenario is that, with the passing of time, the superficial rocks were increasingly oxidized and then transported to the mantle through the process of subduction, which we will discuss in greater detail in the next chapter. During subduction, the rocks in the oceanic crust descend to great depths, melting at least partially and melding with the mantle. In this way, the mantle was slowly enriched with oxygen until it went from being reducing to a more neutral situation. After this process, which was hypothetically concluded around 2.4 million years ago, the gases emitted by the volcanoes were less reducing, and the oxygen was finally free to accumulate in the atmosphere.

We do not actually know whether this interpretation is correct or not, and it is almost certain that the entire process of oxygenation would have been much more complicated, influenced by many other factors and difficult to reconstruct. What we do know is that, over a period of many hundreds of millions of years, the concentration of oxygen in the atmosphere rose from the negligible levels of the Archean to levels that are comparable (though inferior) to what they are currently. The world had changed forever. Or, at least, until plant life began to thrive on Earth.

The Ozone Layer

Together with the growth in atmospheric oxygen, the concentration of ozone in the stratosphere also increased. Ozone is, in fact, produced through the 'photodissociation' of oxygen molecules (O_2, 2 oxygen atoms) by sunlight. Atoms of free oxygen then combine with other substances, but some combine instead with another molecule of diatomic oxygen, forming a new molecule with 3 oxygen atoms (O_3) and known as ozone.

The ozone that accumulates in the stratosphere (the atmospheric layer between around 15 and 50 kilometres from the ground) plays a crucial role for life on Earth. It is able to absorb the Sun's ultraviolet rays, thus screening the planet's surface from potentially damaging radiation. The majority of organisms are, indeed, vulnerable to ultraviolet rays (just think of the UV lamps used to sterilize medical and surgical instruments) and would not otherwise be able to live on the planet's surface.

It was probably at the same time as this rise in the concentration of oxygen, around 2.4 billion years ago, that the Earth acquired its protective layer of ozone. Since then, it has remained throughout the stratosphere, protecting plants, animals and human beings from the effects of ultraviolet radiation. Ozone, however, is easily destroyed by the chemical reactions caused by compounds such as chlorofluorocarbons.

These are the infamous CFCs and HCFCs whose emission into the atmosphere as a result of human activity caused a 'hole' in the ozone layer some decades ago. This thinning of the layer of stratospheric ozone was particularly evident above the Antarctic. Through a chain of photochemical reactions, any chlorine atom contained in a CFC molecule can destroy many thousands of ozone molecules, making these compounds extremely harmful even at low concentrations. The reduction in the use of CFCs and HCFCs, later regulated through more rigorous legislation, finally led to a substantial reconstitution of the ozone layer.

The Faint Young Sun and the End of the Archean Era

But what was the average surface temperature during the Archean? This is difficult to say as we have very little to go on. Estimates vary from 10 to 80 °C. More recent results, once more based on the analysis of the composition of the oldest rocks, suggest a less extreme climate, with an average surface temperature lower than 50 °C.

These results, however, conflict with what we know about the evolution of the stars, which suggests that the young Sun emitted less energy than it does today. In the 1950s, astrophysicists such as the German Martin Schwarzschild, who later emigrated to the United States, described how the life of a star is marked by different phases. During the first phase, its luminosity (the energy given off by the star) grows over time, and the Sun is no exception. At the time the Earth was formed, its luminosity was just 70 per cent of what it is currently and it has increased gradually to reach today's levels.

With less solar energy reaching our planet, the climate of its first 2 billion years should have been extremely cold, with the Earth completely covered in ice. And yet the results provided by the analysis of Archean rocks and zircons more than 4 billion years old tell a different story. The ancient zircons found in the Jack Hills in Australia were formed at depth, crystallizing from magmas that came from the melting of matter found in the Earth's crust. The isotopic composition of zircons reflects that of the magmas from which they were formed, which in turn preserve at least some of the characteristics of the original crust matter. And yet the composition of zircons reveals that the crust from which the magmas were formed had previously come into contact with liquid water, presumably from an ocean, in a world that could not have been completely frozen. Something doesn't add up.

This problem, called the paradox of the Faint Young Sun, is first described by Carl Sagan and George Mullen in 1972. Carl Sagan was an excellent astrophysicist and exceptionally gifted scientific educator. He was involved in the golden era of space exploration, also contributing to the Voyager missions, probes that left the solar system carrying information about humanity for other possible inhabitants of the cosmos.

The most obvious way to escape this paradox (as proposed by Sagan and Mullen) is to hypothesize that the primordial Earth had an atmosphere that was extremely rich in carbon dioxide and methane produced by bacteria and capable of trapping infra-red radiation emitted by the Earth, thus generating a greenhouse effect that would have been much more intense than that we see today. This highlights carbon dioxide's staggering importance to the planet's climate. These ancient high concentrations of CO_2 and methane provided our planet with a thick 'blanket' that allowed it to maintain temperatures suited to the presence of life despite weaker solar radiation.

However, towards the end of the Archean, the rise in the concentration of oxygen shocked not only the ecosystems but also the planet's climate. The oxygen reduced the time methane remained in the atmosphere, while also limiting the distribution of those anaerobic organisms (bacteria and Archaea that live in anoxic conditions) known as methanogens due to their capacity to produce methane through metabolism. In the newly emergent kingdom of oxygen, the atmospheric concentration of methane was therefore reduced. But as the Sun was still faint, the Earth found itself in an unexpected position. For perhaps the first time ever, an epochal change in climate occurred and the planet was covered by a thick layer of ice. Snowball Earth was born.

2

A World of Fire and Ice

Ice Everywhere (or Just About)

The probe descended slowly, sending images and data to the mothership orbiting above, beyond the atmosphere of that white planet. On the surface below, ice covered just about everything. Only the peaks of the highest mountains emerged from the dazzling blanket that reflected most of the light received from the star. At the equator, perhaps, the ice opened up leaving a space for the dark ocean waters, which could only have any direct contact with the air at that point and managed to absorb a little light and oxygen, even if we still do not know the actual size of the ice-free area. And yet those same oceans nevertheless hosted a wealth of life – predominantly single-celled organisms, but also more complex, multi-cellular beings.

This is how our planet would have looked to hypothetical space explorers some 600 million years ago. It was Snowball Earth, a climate state characterized by an extensive and thick layer of ice covering almost the entire surface of the Earth. Between the end of the Archean almost 2.5 billion years ago and the beginning of the Cambrian era some 542 million years ago came the 2 billion years of the Proterozoic. During this eon, the Earth went through devastating climate changes, alternating between eras in which ice covered most of the planet and those in which the climate was temperate, or even hot.

The first episode of planetary glaciation probably happened at the beginning of the Proterozoic, shortly after oxygen had begun to accumulate in the atmosphere. This is known as the Huronian glaciation, discovered in 1907 by Arthur Coleman, who was analysing rocks close to the mineral township of Cobalt, in Ontario, Canada. Coleman noticed that enormous and very ancient masses were embedded in the region's geological strata, and that these masses had been transported a great distance from their area of geological origin. They presented all the signs of glacial action, such as smoothed and striated surfaces, similar to those found on the erratic masses of the last, most recent glacial periods. And so, one of our planet's oldest glaciations was discovered.

Similar evidence was recently found in the rocks of many other regions of North America, but also in South Africa, India and Australia. At the time of the Huronian glaciation, the continents were arranged in a different way from today, but this geographical extension suggests that ice covered most of the Earth, and not just for a brief time. The duration of the Huronian glaciation was around 300 million years, a period in which many species of living organisms became extinct.

Is It Always Oxygen's Fault?

The obvious question is, naturally, why a glaciation of this scale happened. Which mechanisms caused the planet to fall into this icy grip? The most likely hypothesis (though not the only one) is once again associated with oxygen and the enormous upheaval this gas brought to the planet.

As we have seen, it is highly likely that methane was a key element in the atmosphere's composition, responsible for a particularly intense greenhouse effect and capable of countering the scarce solar luminosity of that time, meaning surface temperatures would continue to be mild or even elevated. But the arrival of oxygen threw a spanner in the works. Oxygen reacted easily with methane, eliminating it from the atmosphere and drastically reducing the greenhouse effect, which caused a subsequent crash in temperature. This is the chain of events that probably led to the first great planetary glaciation.

Oxygen's responsibility for triggering a glaciation was first identified by Canadian geologist S. M. Roscoe, once again studying the rocks around Lake Huron, not far from where Coleman had found proof of the Huronian glaciation. Roscoe noted that, just below the blocks striated by the glaciation, there were strata that contained detritus of pyrite and uraninite, minerals that can only be deposited in sediment in conditions of oxygen scarcity. Just above the strata of glacial rocks, there was instead evidence of oxidized sediment, a sign that oxygen was present in the atmosphere. This observation led to the hypothesis that it was precisely this rise in atmospheric oxygen that reduced the concentration of methane and therefore the greenhouse effect, unleashing the glaciation.

In this case, we also see how the accumulation of oxygen, due to the actions of photosynthetic organisms, caused irreversible changes in the planet's climate and environment. As the ice started to expand, the planet's surface became increasingly white and reflective. This meant less sunlight was absorbed, which led to a further drop in temperature. Around 2.4 billion years ago, therefore, our planet first ran the risk of falling into a state unsuited to life. But if we look at the Earth today, it is clear that it managed to find a way out.

The Volcanoes Set the Planet Back in Motion

We might then ask ourselves how our planet managed to escape the ice's frozen clutches. One possibility is linked to volcanic activity. In the Proterozoic, the volcanoes were extremely active and the eruptions more violent and frequent than those we see today. On the land that emerged, there were no expanses of forest or prairie. The landscapes were barren and rocky, the organic soils produced by the actions of living organisms had yet to form, and the wind and rain swept across continents inhabited solely by bacteria in the areas with the most water, along streams and rivers, in swamps and around hydrothermal areas.

With the glaciation, the emergent lands were also covered (at least partially) by glaciers. The blanket of ice and the low temperatures slowed down the processes of surface rock erosion and weathering (physical and chemical alteration), the main geological mechanism that removes carbon dioxide from the atmosphere, as we shall discuss further in due course. But carbon dioxide was still emitted by volcanoes, which filled the atmosphere with large quantities of the gas. Even without the methane, which had by then been overcome by the oxygen, over the course of millions of years it is probable that the concentration of carbon dioxide continued to rise, increasing the greenhouse effect until it reached a critical threshold that caused temperatures high enough to melt the ice sheets and bring about the end of the Huronian glaciation. The entire melting process may well have been more complicated – aided, for example, by the cloud cover which was probably present even above the ice. In any case, just over 2 billion years ago, the Earth escaped this icy grip and a new era began. For a long period of time, almost 2 billion years, the planet would have been hot, with little or no ice coverage, and with more and more oxygen in the atmosphere.

Emerging Rocks

There were no particularly significant planetary events in the billion years that followed the end of the Huronian glaciation, leading to it being nicknamed the *Boring Billion*. However, during this period, other aspects of the planet were very active indeed, such as the shifting of the continents due to movement deep in the Earth's mantle, the birth of a new continental crust, and plate tectonics.

Our planet's surface is formed by the Earth's crust, of which there are two kinds: the continental crust, which is mostly made up of granite, and the oceanic crust, primarily composed of basalt, which is denser than granite. The Earth's mantle is formed from even denser rocks and has a temperature that rises the deeper it goes, to such an extent that the rocks in the mantle are rendered malleable and, in some parts, partially molten. The crust varies in thickness from 5 kilometres at the oceanic crust to 70 kilometres under the highest mountain ranges of the continental crust, whilst the mantle has a thickness of around 2,900 kilometres. If we descend even farther, we come to the realm of the core, composed predominantly of molten metals on the outside and, at its centre, solidified metals rendered extremely dense by the enormous pressure. At the heart of our planet, therefore, is an enormous solid block of metal covered by an ocean of molten metal in constant motion. Truly a scene worthy of Jules Verne.

In the mantle are radioactive elements, trapped there since the Earth's formation, which decay slowly over time. This decay generates the heat that warms the mantle from the inside, causing the very slow movement and deformation of rocks. This movement is much like the circulation of water in a saucepan when heated from below. The hot liquid rises to the surface where it releases heat and slowly returns to the bottom, creating circulation loops known as 'convection cells'. The heated air moves in the same way, generating sometimes violent phenomena such as storms.

Convection cells in the mantle obviously move more slowly than those in the air, and are more akin to the moulding of a block of plasticine than to the circulation of a liquid. A full rotation on this merry-go-round, from the depths of the mantle to the surface and back again, takes between 50 and 200 million years. The entirety of the Earth's surface is subdivided into plates, very large regions that correspond to the enormous convective cells in the mantle, which, together, move the rocks and carry heat generated in the depths up to the surface. These movements are described today by plate tectonics, the modern incarnation of the theory of continental drift first proposed (though unappreciated at the time) by Alfred Wegener, a German geologist and meteorologist who spent a long time studying in the Arctic (today, Germany's institute for marine and polar research is named after him). Plate tectonics is a crucial process for our planet that was only discovered in the 1970s and is still not fully understood today. It is curious to think that, decades after the theory of relativity, after quantum mechanics and space travel, we still had no idea how the inside of the Earth was structured; nor were we familiar with the most fundamental geological mechanism characterizing our planet.

In the ascending branch of convection cells, the lightest and partially molten components of the rocks in the mantle rise to the surface, accompanied by the formation of volcanic systems and eruptions of magma that, when they solidify, generate basalt, the black rocks of the oceanic crust. In the middle of the Atlantic, roughly parallel to the American and African coasts, a huge underwater mountain range signals the presence of the ascending branch of a convection cell. Similar mid-ocean ridges can be found in the Pacific, the Indian Ocean and in many other areas of our planet. The Mid-Atlantic Ridge runs through Iceland, creating the volcanic landscape found on that island of fire and ice. In Ethiopia, in the largest depression of the Rift Valley, the African plate is breaking up, moving away from the Arabian plate and creating the basis for a new ocean that will see the Red Sea grow much larger.

During the volcanic processes associated with convection in the mantle and movement in the plates, enormous quantities of carbon dioxide and other gases are emitted, which accumulate in the atmosphere and strengthen the greenhouse effect, favouring a rise in surface temperatures. If there were no efficient mechanisms for removing these gases, the temperature would keep rising, and the water in the oceans would evaporate up to the high atmosphere where it would be broken down into oxygen and hydrogen. The hydrogen would then be lost, as it can easily escape the planet's gravity, and the result would be an out-of-control greenhouse effect as happened on Venus, where surface temperatures exceed 400 °C.

How the Planet Keeps the Greenhouse Effect at Bay

The carbon dioxide that builds in the atmosphere due to volcanic activity is removed by chemical and geological processes that take place on the continental areas and in the shallow seas that surround them. The first step involves precipitation. Carbon dioxide combines with the water in raindrops and forms the weak carbonic acid that chemically alters the rock as it falls onto the Earth's surface, dissolving it and freeing ions of magnesium, calcium, sodium and potassium. The rivers then transport these elements to coastal areas. Here, the calcium and magnesium ions combine with the bicarbonate dissolved in the water, producing calcium carbonate (the limescale in drinking water that clogs up taps and kettles) or magnesium carbonate.

In the sea, the carbonate is predominantly used by planktonic organisms such as foraminifera to build protective casings (much like a shell but on a microscopic scale). When these organisms die, the now-empty casings fall slowly to the seabed. This process leads to an accumulation of calcium and magnesium carbonate, particularly on the continental shelf surrounding the coastlines. Over thousands of years, the small shells bond together and form large deposits of carbonate rocks, as we can see, for example, with the white cliffs of Dover in England. With the movement of continents and the disturbance of geological strata, these deposits can go on to form spectacular mountains or, if subjected to very high pressure, be transformed into marble.

The clear effect of this weathering process on the rocks is that carbon dioxide is removed from the atmosphere and carbon is stored in the seafloor. The chemical and geological process is accelerated significantly by the action of living organisms, both during the actual weathering of the rocks on the continents, and when the calcium carbonate is deposited in the sea.

If there were no other processes producing carbon dioxide, this gas would be extracted from the atmosphere and the greenhouse effect would diminish, leading once more to a state of widespread glaciation.

We have looked at two competing processes: the emission of carbon dioxide into the atmosphere by volcanoes, and its elimination by processes of weathering. But how can volcanoes continue to produce carbon dioxide? Where do they get it from?

Rocks that Sink

The key to closing the geological carbon cycle, and therefore that of carbon dioxide, lies precisely in plate tectonics. In the most external part of the solid Earth, the continents (which are lighter) float on the mantle and the oceanic crust, and are shifted by convection movements as if they were corks in a pan of boiling water. But, unlike corks, the continents deform and when the mantle's convective motions cause them to meet, they fray, buckling against one another, mixing together until they become one. And so, mountain ranges are born, their original components often still distinguishable in the chaos of overlapping rocks. This is how the Alps were formed, just like the Rocky Mountains, the Andes and the Himalayas.

If a denser component of oceanic crust is present when plates collide, this is drawn down into the mantle in a geological process known as subduction, which we discussed in the first chapter. In the mantle, the extreme pressure and high temperatures transform the sedimentary rocks into metamorphic rocks, freeing carbon, oxygen, hydrogen and other volatile elements present in the minerals in various forms. In this way, part of the carbon previously contained in the calcareous rocks or in the sediments rich in organic matter is freed, forming carbon dioxide once more. The fluids generated in this way find a route upwards, sometimes through fissures and cracks (known as faults), sometimes transported by the magma in the volcanoes. The overall effect is that the carbon dioxide once fixed in the weathered rocks is now regenerated and released into the atmosphere once more.

Subduction brings the geological cycle of carbon to a close, from its emission by volcanoes to its precipitation to the surface in rain, and its role in the weathering of silicate rocks; the creation of carbonate casings by plankton and their subsequent sinking to the seabed, where they give rise to carbonate rocks that then sink farther through subduction, releasing carbon dioxide; and, finally, the emission of this gas into the atmosphere by volcanoes that develop in areas where the plates collide. The entire geological cycle of carbon takes between 100 and 200 million years, a time period that is roughly the same as that taken to complete a full cycle of convection in the mantle. Superimposed on and interacting with the geological carbon cycle is the biological carbon cycle (photosynthesis and respiration), with its greater flows and shorter time scales, but with a much smaller total quantity of carbon at play than that in the mantle.

It is curious to note that, today, the only planet on which we know plate tectonics are present is Earth. Venus, Mars and Mercury do not currently have global plate movements like those seen on Earth. We might ask ourselves, then, whether it actually was plate tectonics, with its processes of the subduction of rocks and recycling of gases, that rendered the concentration of carbon dioxide in our atmosphere relatively stable, maintaining the temperatures at a level compatible with the presence of liquid water and, ultimately, allowing for the permanence of life on Earth. Currently, we can neither prove nor deny this, but in the coming decades, with new observations of planets outside our solar system, we will perhaps be able to understand better whether it is this geological process in particular that makes our planet habitable. There are also those who believe that the search for 'habitable' planets (those that allow for the presence of life, though not necessarily our own) actually means looking for rocky planets characterized by active plate tectonics.

Ice in Tropical Seas

At the end of this period of relative stability, just over 700 million years ago, the climate situation changed radically. From this almost tropical world, free (or just about) of ice, a new frozen environment was about to form. For some dozens of millions of years, the ice pushed towards the equator, creating that white planet we fantasized about at the beginning of the chapter. This moment in the history of the Earth is symbolic of how the planet's climate dynamics can generate extreme conditions very different from the current situation or that which has directly preceded it. Greater temporal proximity to this last global ice age has allowed researchers to observe and gather much more information than on the Huronian glaciation, information that also tells us how a planet's climate functions and how it can potentially develop unexpected instability.

Since James Thomson's 1971 work on the Isle of Islay (in the Scottish Hebrides), geologists have gathered more and more evidence of glacial traces dating back to the late Proterozoic, later identified as having originated at tropical and equatorial latitudes. And so, the idea began to take shape that in a timeframe somewhere between 600 and 700 million years ago, the Earth was completely covered (or almost) by ice.

In 1949, the idea of a global glaciation was clearly expressed by Sir Douglas Mawson, Australian geologist and polar explorer, following his study of rocks of glacial origin in the Elatina Formation in southern Australia. However, his idea was forgotten over the following years, not least because the ground-breaking development of plate tectonics indicated that the positions of the continents have changed drastically over the course of millions of years. In 1964, however, Walter Brian Harland, a geologist at Cambridge University, revived the idea of global glaciation. Harland analysed the magnetic characteristics of rocks of glacial origin (tillites) collected in Greenland and on Svalbard, suggesting that, when they were formed, those rocks (and those islands) were located in the tropics. Indeed, when a rock begins to form from either sediment or magma, crystals of ferrous minerals orient themselves in the direction of the Earth's magnetic field. By studying the orientation of the field that has remained fixed in the rocks, it is possible to ascertain the approximate latitude of the place they were formed. Tillites in particular are generated through the compression of fine sediments produced by the movement of glaciers, and Harland's discovery therefore indicated the presence of glaciers in the tropics!

In 1987, geochemist Joseph Kirschvink and his collaborators once more studied the sedimentary rocks of glacial origin in the Elatina Formation, using a more refined analysis of their magnetization to demonstrate clearly that they were formed at subtropical latitudes and close to the sea. This observation therefore confirmed the presence of glaciers at sea level in tropical and equatorial areas. It was Kirschvink himself who coined the expression Snowball Earth in an article published in 1992.

Over the following years, the work of Paul F. Hoffman and Daniel P. Schrag, together with that of many other researchers, shed light on more and more details from this particularly special period in the history of the Earth's climate. Even today, it is not yet clear whether Snowball Earth was characterized by a total covering of ice all the way to the equator, or whether there was a large strip of equatorial ocean that remained free of ice, as hypothesized by William Hyde, Thomas Crowley and collaborators. In any case, ice covered the vast majority of the Earth at least twice, each for a time period spanning many millions of years, in a timeframe dating from around 720 to around 635 million years ago.

But what caused such a change in the planet's climate, allowing for such an extensive covering of ice? If oxygen was responsible for the first Huronian glaciation, what caused the global glaciation at the end of the Proterozoic? Kirschvink himself proposed an explanation that continues to be substantially valid today. Due to movement deep in the mantle, a little more than 700 million years ago, the supercontinent called Rodinia, which accounted for most of the continental lands, began to fragment, generating a multitude of large and small islands with broad areas of continental shelf and low seas. Furthermore, many of these continental fragments found themselves close to the equator by chance, as we can confirm through geological analysis. Now, the mechanism of surface rock weathering is strongly dependent on temperature and precipitation. The concentration of continental masses close to the equator, with large areas of shallow surrounding seas, thus accelerated weathering processes in the rocks and the sedimentation of carbonates, removing the carbon dioxide from the atmosphere more effectively and drastically reducing the greenhouse effect. The absence of ice at the poles also exposed more areas to weathering, contributing to the planet's plunging temperatures.

In addition to this, in the Proterozoic the continents were still free of vegetation. As bare rock tends to be lighter than the ocean, it reflects more of the incident solar rays. It is precisely at the equator that the most solar radiation is received, and the reflection effect therefore has the greatest consequences. The combination of diminishing carbon dioxide and a lower absorption of sunlight causes temperatures to drop, leading to a widespread covering of ice, first at the poles and then gradually at lower latitudes. The farther the ice expands, the more sunlight is reflected, the more the temperatures drop. Within a few tens or hundreds of thousands of years, the Earth was once again in an icy grip.

A last hypothesis with which to explain such a change in climate is, once again, oxygen. Shortly before the Snowball events, the atmospheric concentration of oxygen increased further, perhaps due to an imbalance in the production of oxygen through photosynthesis and its consumption in the decomposition of organic matter. While the presence of methane in the Proterozoic atmosphere was still significant (despite being at a lower level than in the Archean), the rise in oxygen may have caused it to disappear altogether, thus rapidly reducing the greenhouse effect and contributing to the cooling.

During the Snowball, the oceans covered in ice and with little contact with the air became, for the most part, anoxic, despite the greater concentration of oxygen in the atmosphere. The iron began to accumulate in suspension once more, as had already happened billions of years earlier. The ice sometimes melted and the ocean was once again partially in contact with the atmosphere, and this injection of atmospheric oxygen caused the iron to sink, generating for the last time those red rocks of the banded iron formations seen a few billion years earlier.

Even during the extreme conditions of the global glaciation, life on the Earth did not become extinct. Bacteria and, for the most part, prokaryotes (organisms without a distinct cellular nucleus) continued to exist in large quantities, as well as eukaryotes (organisms whose cells have a separate nucleus) and even the multi-cellular organisms that had begun to evolve before the advent of the Snowball, all managed to survive, albeit with obvious difficulty. Understanding how the eukaryotes had been able to continue living in such extreme conditions remains complicated. However, there were obviously some environments that sustained the existence of living multi-cellular organisms, such as the partially defrosted areas within the expanses of ice, temporary apertures in the sea ice (which are still present in the polar pack ice and are known as polynya), swamps of meltwater and, above all, hydrothermal vents in shallow waters. Another possibility is that on the edge of the continents, where the glaciers entered the ocean (as happens today in the polar fjords), the meltwater was sufficiently rich in both oxygen and nutrients to meet the needs of those living organisms. Furthermore, algae can live close to the ice's surface (as happens today in mountain glaciers), managing at least partially to utilize sunlight for photosynthesis. Of course, if the equatorial oceans remained free of ice, it would have been easier for those living organisms to survive in that equatorial band exposed directly to the sunlight and exchanges with the atmosphere. But beyond these details, which are obviously central from a scientific perspective, it was, in any case, a fascinating world, a primordial and unwelcoming 'Realm of Winter' that lasted for tens of millions of years. It is difficult even to imagine how such a world might have been, so different from the blue and green planet we know today.

Volcanoes and the Start of a New World

At a certain point, the last Snowball also ended. It is likely that, again, in this case it was the volcanoes that led to the ice melting – volcanoes that emerged from the blankets of ice, more or less as Vatnajökull in Iceland appears today. Without any superficial weathering of the rocks, the carbon dioxide accumulated in the atmosphere, reaching much higher levels even than those of the following eras. In order to deliver a world from perennial ice, however, it is necessary to pump a truly massive amount of carbon dioxide into the atmosphere. And perhaps this is what the volcanoes did. The greenhouse effect set to work, the temperatures rose, the ice melted, and the snowball turned into water. The eternal winter had ended. Indeed, the high levels of carbon dioxide generated tropical conditions throughout the globe with high temperatures, heavy rain, rapid sedimentation, and the flourishing development of life. Today, evidence of this period of impetuous heat can be found in the thick deposits of the carbonate strata that are superimposed onto the sediments deposited during the glacial period.

From this point on, there were no further episodes of such extensive ice coverage, though glaciations of more modest proportions continued to take place. There may be several reasons why no more global glaciations occurred. Firstly, continents were no longer amassed at the equator, and secondly (though no less importantly), prairies and forests had developed on the continents, reducing the bare rock's reflective capacity and causing the climate to stabilize, as we will see in the following chapters. Finally, while the definitive disappearance of methane from the atmosphere had played a role in triggering the global glaciations, since the last Snowball the greenhouse effect has been dominated by carbon dioxide, which has rather different mechanisms of production and removal.

27

Immediately after the end of the Snowball, around 635 million years ago, life on Earth underwent a sudden development characterized by the biological evolution of a vast number of new multi-cellular organisms, some of which provided the origins for our own species. These few dozen million years would radically and permanently change the history of the Earth from the 4 billion years that had gone before.

3

Light Reflected, Light Re-radiated

To understand how a planet's climate works, we must identify the mechanisms that determine its characteristics and variability. In the two previous chapters, we discussed how snow and ice reflect a large part of the sunlight, and how variations in greenhouse gas concentration (primarily carbon dioxide, and methane in more remote eras) can cause enormous changes in planetary temperatures. Light absorbed and light re-radiated: these are two of the main protagonists when it comes to the Earth's climate.

Children of the Sun

The great engine behind our planet's climate is the Sun. If our star were to go out, the Earth would become dark, cold, covered in perennial ice that could well have a slightly melted base due to the geothermic heat coming from the depths of the mantle. It would look like those planets ripped from their star system that, it is assumed, are able to wander through the vastness of space, ghost ships that perhaps once teemed with life.

Luckily for us, our Sun shines brightly and is (for now at least) stable, heating the planets that rotate around it. The solar energy that reaches the Earth every second (the energy per unit of time is called power) is measured today by satellites, and is called the solar constant. The solar constant measures the radiation that hits a surface placed perpendicular to the incident solar rays, situated at the average distance of the Earth from the Sun. The Sun actually has a cycle of luminosity variation that lasts around 11 years – visible also from the number of sunspots – during which there are mild variations in the power emitted. The solar constant, which we will refer to as S, is estimated to be just under 1,361 W per square metre (W/m^2) during a minimum of solar activity, and around 1,362 W/m^2 during a solar maximum, meaning there is a variation of around 0.1%. Over longer time periods of some hundreds of years, it is possible that there might be greater variation in the energy emitted by the Sun. For example, the unusually cold period experienced by many parts of the world between the end of the 1300s and the middle of the 1800s, known as the Little Ice Age and which we will discuss later on, has sometimes been attributed to reduced solar emissions. In any case, analysis of the available data indicates that, over the last 400 years, solar luminosity has not varied more than 0.2%.

As we have said, the solar constant measures the power that arrives from the Sun at the average distance of our planet from its star. Indeed, the power that reaches a surface that is perpendicular to the incident solar rays decreases with distance from the Sun. The farther away we are, the less solar energy we receive. A lot less! For example, Mars receives on average less than half the power that reaches the Earth, and Pluto, beyond Neptune's orbit, receives less than a thousandth of the power that reaches our planet.

As the orbit of the Earth is slightly elliptical, as it moves our planet can find itself a little closer to the Sun (during summer in the Southern hemisphere) or a little farther away (during summer in the Northern hemisphere). As a consequence, the solar power reaching the Earth also varies with the seasons. Significantly, the power arriving at the two extremes of the orbit (the point closest to the Sun, called the perihelion, and that farthest away, the aphelion) varies to almost 3% more or less than the average.

To complete the picture, we must consider another relevant fact: the Earth is round. We have known this for thousands of years, but we sometimes forget. We have learnt since childhood that the poles are cold because the Sun's rays arrive there "at a greater incline" and therefore their energy is distributed over a larger surface. The same thing happens at both sunrise and sunset, while during the night we receive absolutely no sunlight whatsoever. This means that, averaging out over an entire day, the solar energy reaching the Earth is distributed over the entire surface of the planet. The power would be S multiplied by the planet's great circle, πR^2, in which R is the radius of the Earth, R = 6371 km, and π is the Greek symbol denoting the number 3.14159265. But this power is distributed throughout the entire surface of the sphere over 24 hours, which has a value of $4\pi R^2$. Therefore, in order to calculate the power per square metre that the Earth receives on average over a 24-hour period, we must divide the total power received ($\pi R^2 S$) by the planet's surface area ($4\pi R^2$). The result is $S/4$, or just over 340 W/m^2 (calculated here at the solar minimum). The value of $S/4$ is therefore the 'number' that acts as the starting point for our considerations of the climate.

Shining with Reflected Light

In the cosmos, there is a fundamental difference between stars and planets (not to mention satellites, asteroids and so on). If we consider the visible portion of the electromagnetic radiation spectrum known as the visible light spectrum, with a wavelength between around 390 and 700 nanometres (1 nanometre is 1 billionth of a metre), the stars shine with their own light, which is produced by the nuclear fusion reactions that take place in the extremely compressed matter that makes up the star. Planets and satellites, however, reflect only the visible light that reaches them from elsewhere, such as from their star. This is how we can see the Moon, Mars, Venus and all the other bodies in our solar system. The Earth also reflects part of the light it receives from the Sun, faintly illuminating the unlit part of the Moon (known as Earthshine). In this case, the light of the Sun hits the Earth, which reflects part of this light towards the Moon, which in turn reflects it back towards the Earth causing the light to diminish in intensity.

But how much light is actually reflected from a planet or a satellite? Let's look at a familiar case: Earth. If we leave two objects in the sun, one light and one dark, we can see that the dark object heats up much more than the light one. This is because light objects reflect most of the sunlight that hits them, whilst darker objects tend to absorb it. The amount of incident sunlight reflected depends, therefore, on the colour and type of surface. Light sand, for example, reflects most of the light it receives, as do snow, ice and clouds. Ocean water and forests, on the other hand, reflect much less and absorb most of the incident radiation.

On average, the Earth reflects around 30% of the incident solar radiation that reaches it. The fraction of reflected radiation is called albedo, and the average albedo of our planet is 0.3. For the most part, this value is due to the bright polar ice sheets, and in particular to the clouds that cover more than 60% of the planet's surface at any given moment.

The calculations that we have just looked at must, therefore, be reconsidered. Because, if it is true that the average solar power received by the Earth is S/4 (which, as we have said, is just under 340 W/m²), it is also true that part of this radiation is reflected into space and does not contribute to the heating of the planet's surface. If we remove the 30% that is reflected, the power absorbed therefore becomes about 238 W/m². This is the value we must use to calculate an initial energy budget of the Earth's climate system.

Energy Absorbed and Re-radiated

We have just seen that, over the course of a day, the Earth absorbs on average around 238 W/m², and, as a consequence, it heats up. If there were no mechanism to free the planet of this heat, the temperature would just keep rising to the point that the rocks would melt and evaporate, transforming the Earth into a ball of gas.

Luckily for us, such a mechanism does exist. Any body with a temperature above absolute zero (zero degrees Kelvin corresponds to –273.15 °C) emits energy. The total power emitted (in watts per square metre) is expressed in an ideal situation using the formula σT^4, in which σ is a constant (named the Stefan–Boltzmann constant after the two great Austrian physicists Josef Stefan and Ludwig Boltzmann who studied this process at the end of the nineteenth century), and T is the temperature in Kelvin of the body's surface. Therefore, the hotter a body, the greater the power emitted, with a very high dependence (to the power of four) on temperature.

The Earth, heated by the Sun, therefore in turn emits energy, freeing itself of the heat it has absorbed and reaching (if everything remains constant) a state of equilibrium in which the power absorbed is equal to the power emitted. Bearing in mind that the power absorbed is worth (1–A) S/4, in which A is the planet's albedo, the simplest climate model can be expressed as power emitted = power absorbed, or:

$$\sigma T^4 = (1-A) \cdot S/4$$

From this formula, we can estimate the average temperature of our planet. If we insert the correct values of S, A and the Stefan–Boltzmann constant σ, we find that the Earth's average temperature should be around 255 Kelvin – that is, –18 °C. As the average temperature of the Earth's surface is actually +15 °C, it is clear that we have made a mistake, or forgotten something.

The Earth's Blanket of Gas

A heated body re-radiates the absorbed energy, and the higher the temperature, the more intense the re-radiation. The energy is emitted in the form of electromagnetic waves of varying lengths. Radio waves have a longer wavelength than infra-red, which in turn have a longer wavelength than visible light, whilst, for example, X-rays have an even shorter wavelength again. The distribution of energy emitted at varying wavelengths from an ideal body is called blackbody radiation, and was calculated exactly by physicist Max Planck in 1900. What interests us here is the fact that a large part of the power emitted is concentrated at a specific wavelength, λ, which gets shorter as the temperature rises. This relationship, which can be written as $\lambda \cdot T$ = constant, is called Wien's displacement law.

When using this law, we discover that the dominant wavelengths emitted by a body with a surface temperature of around 6,000 Kelvin (like the external part of the Sun, the source of the light that reaches the Earth) correspond to the visible part of the electromagnetic radiation spectrum. Whereas a body with a surface temperature of around 290 Kelvin emits predominantly infra-red radiation. Earth, therefore, emits infra-red radiation. Our eyes are not capable of seeing these wavelengths (hence why we need special binoculars, such as those used by security forces), and so we do not see the light radiated by the Earth, but only the visible sunlight the Earth reflects. But if we were to look at Earth from space with an infra-red sensor, we would see it shine not just with reflected light, but also with re-radiated light.

It is precisely the difference between absorbed visible light and re-radiated infra-red radiation that explains what we have forgotten in our recent calculation. If the Earth had no atmosphere, the estimated temperature of –18 °C would be substantially correct. However, the 33 degrees of difference between the theoretical calculation and measurements are due entirely to the fact that the Earth is enveloped in the gassy blanket of its own atmosphere.

Around 78% of the Earth's current atmosphere is composed of nitrogen, an inert gas transparent to both visible and infra-red radiation. Just under 21% is oxygen, also transparent, and a little less than 1% is argon, another inert gas. The tiny remaining fraction is comprised of both water vapour and gases such as carbon dioxide and methane, referred to collectively as greenhouse gases.

Let's look first at carbon dioxide, the average atmospheric concentration of which is today less than 420 parts per million (ppm, which means that in a cubic metre of air there are 420 cubic centimetres of carbon dioxide), but which is undergoing constant and rapid growth as a result of emissions of anthropogenic origin. As we have seen in the previous chapters, atmospheric carbon dioxide has played a key role in determining our planet's climate. Before this, it was methane that generated the greenhouse effect that kept the temperatures high despite the presence of a faint young Sun. This is because the greenhouse gases (carbon dioxide and methane) and water vapour are pretty much transparent to the light that comes from the Sun, but they absorb infra-red radiation very effectively. Greenhouse gas molecules in particular can be set in rotation, vibration or deformation by radiation with a wavelength inside the infra-red range of the spectrum.

This is why the infra-red radiation emitted by the Earth's surface does not immediately escape into cosmic space, but is partially absorbed by the layer of greenhouse gases present in the atmosphere. Planets such as Mercury and satellites such as the Moon, which lack a sufficiently dense atmosphere, cannot produce this effect and the infra-red radiation emitted by their surface is dispersed in space.

The molecules of greenhouse gases, absorbing part of the infra-red radiation, reach an energetically excited state before returning to their original condition and re-emitting the infra-red radiation. However, only around half of the radiation absorbed is emitted upwards, while the other half moves towards the Earth's surface, where it is absorbed once more. If ε refers to the fraction of infra-red radiation emitted by the surface and absorbed by the greenhouse gases in the atmosphere, then the fraction re-radiated towards the Earth would be $\varepsilon/2$. Therefore, the surface of the planet will receive not only solar radiation – the value $(1-A) S/4$ from our previous formula – but also the quantity $\varepsilon/2 \; \sigma T^4$ that the greenhouse gases send downwards. Our climate model should therefore be written as:

$$\sigma T^4 = (1-A)\ S/4 + \varepsilon/2\ \sigma T^4$$

or

$$\sigma T^4 = \frac{(1-A)\ S/4}{1-\varepsilon/2}$$

The overall result is that the heat balance at the Earth's surface will lead to a higher temperature than we had previously calculated, and the greater the value of ε, the higher it will be. The value of ε, in turn, increases with the concentration of greenhouse gases in the atmosphere. Just as a blanket stops the dispersion of body heat, our blanket of greenhouse gases stops the heat emitted by our planet being entirely lost in the cosmic cold.

How Is This Blanket Made?

The first quantitative intuitions on the role played by atmospheric carbon dioxide in the Earth's climate were expressed in 1856 by US scientist Eunice Newton Foote, who demonstrated carbon dioxide's capacity to re-radiate the heat it receives. Her contribution and role as a researcher have been almost entirely ignored until today, partly because, shortly after her work was published in 1856, Irish physicist John Tyndall refined the 1827 work of French mathematician and physicist Jean Baptiste Joseph Fourier exploring the same theme. But it was Swedish chemist and physicist Svante Arrhenius who, in 1896, published a fundamental work on the link between carbon dioxide and planetary climate, paving the way for further research and in which he clearly expressed his concern that a rise in atmospheric carbon dioxide due to industrial emissions could impact the climate and its temperatures.

As we have said, the intensity of the greenhouse effect depends on the value of ε. The greater the concentration of carbon dioxide, methane or water vapour, the greater ε will be and the more intense the greenhouse effect. In this regard, we have seen how a considerable drop in the concentration of carbon dioxide unleashed a precipitous dive into the frozen world of Snowball Earth, and how the accumulation in the atmosphere of carbon dioxide emitted by volcanoes eventually caused the Earth to defrost, generating a period of tropical heat.

The rise in temperature at the surface does not, of course, lead to an energy gain out of the blue. The heat trapped beneath the layer of greenhouse gases and the increase in surface temperature come at the expense of the heat that should have been absorbed by the highest layers of the atmosphere. The greenhouse effect, therefore, also causes a drop in temperature in the high atmosphere, which compensates for the greater heat trapped close to the planet's surface.

What are the main greenhouse gases? We have mentioned carbon dioxide (CO_2), methane (CH_4) and water vapour (H_2O). But there is also nitrous oxide (N_2O), sulphur hexafluoride (SF_6), ozone in the troposphere and chlorofluorocarbons (CFCs), the same compounds responsible for the thinning of the ozone layer high in the polar atmosphere. When equal in concentration, methane and water vapour generate a much more intense greenhouse effect than carbon dioxide. Since the atmosphere grew rich in oxygen at the end of the Archean, the concentration of methane has remained relatively low and the effect of carbon dioxide came to dominate. The situation is, however, very different when it comes to water vapour. Vapour is a great amplifier of the greenhouse effect and today plays an essential role, but it is generally passive and adapts to the temperature and pressure conditions in the atmosphere.

In a world like that of the Earth, in which there is no lack of water, vapour rises easily from the surface of the ocean or is transpired by vegetation. But there is a maximum volume of vapour that the air can contain. This maximum value depends heavily on air temperature, according to a law known as the Clausius–Clapeyron law, from the names of German physicist and mathematician Rudolf Clausius and French physicist and engineer Émile Clapeyron who formulated it from their experimental observations. This law states that the higher the temperature, the greater the quantity of water vapour that can be contained in the atmosphere. Therefore, if we suppose that there is a rise in the concentration of carbon dioxide, the increase in the associated greenhouse effect causes a rise in the temperatures of both the surface and the air with which it is in contact. But at higher temperatures, the air is able to hold a greater quantity of water vapour and, given that sources of water are abundant, this will occur due to more intense evaporation favoured by the higher temperatures. And if the quantity of water vapour increases, both the greenhouse effect and the temperatures rise in a mechanism of reciprocal amplification. This is why the role played by water vapour is not so much that of an autonomous driver, but rather one that amplifies the greenhouse effect generated by the carbon dioxide.

We will return to this later on, but for now we can observe that, under certain conditions, this mechanism of amplification can get out of control. The temperature rises, the water evaporates, the greenhouse effect increases, the temperature rises further still, even more water evaporates, it gets even hotter, until the oceans evaporate and surface temperatures reach levels at which the rocks begin to melt. The molecules of water vapour in the atmosphere are split into their two constituent parts – hydrogen and oxygen – by the sun's rays. Hydrogen, which is much lighter, is lost to space because it cannot be held by the gravity of the planet, whilst oxygen chemically binds with the surface, oxidating it. This terrifying process of runaway amplification is what happened on Venus, which has entirely lost its surface water and today has an atmosphere of extremely dense carbon dioxide. This is a very unhospitable world and a climate condition from which it is, unfortunately, very difficult to return.

Albedo Again

Before concluding our rapid journey through the Earth's energy balance, let's return to the albedo and focus on an approximate calculation that will allow us to get a general idea of how important this parameter is. Today, the average temperature of the Earth's surface is around 15 degrees centigrade (Celsius, °C), or, in absolute temperatures, around 288 Kelvin (you can calculate degrees Kelvin by adding 273.15 to °C). If the Earth's albedo were to change by 1%, for example becoming 29% instead of 30%, there would be an additional absorption of sunlight of around 3.4 W/m^2. Using the same formulas as earlier, we find that, in the current conditions of surface temperature, this change would bring about a rise in global temperatures of around 1 °C.

Naturally, this small calculation should be treated with caution, as it ignores the processes that accompany changes in the albedo, from variations in the emission and absorption of greenhouse gases to changes in atmospheric and marine circulation. Furthermore, a change in albedo of around 1% is anything but small, and requires significant modifications in cloud cover or the distribution of ice or vegetation. But, in any case, this result tells us how important the albedo is and the considerable impact any change in the quantity of reflected energy can have: so considerable as to cause different possible climate states without changing the solar radiation received by the planet.

Multiple Balances

If we go back to our simple climate model and suppose that the albedo could be low (as it would in a hot world without ice) or high (as in a cold Snowball), we discover that, with the same quantity of solar power received, our model provides two possible solutions for a stable equilibrium: a situation with little ice, low albedo, lots of absorbed solar radiation and abundant infra-red radiation emissions; and another with extensive ice coverage, high albedo, little solar radiation absorbed and scant infra-red emissions. This result provides an exemplary illustration of how the climate is a complex system capable of manifesting multiple states of equilibrium even when the external forces (in this case, the solar power received) do not change. It is precisely these mechanisms of response and feedback within the climate that make the planet's dynamics so very creative. We mustn't think, therefore, that the climate's variability is the sole result of variations in external forces (such as incident solar radiation). The internal mechanisms can amplify or diminish the variations in these forces, and can generate climate variations even when external conditions are constant. With such dynamic behaviour, it is possible for sudden and unexpected changes to happen even when there are minor variations in external forces, or in response to small changes in the concentration of greenhouse gases.

In the case of Snowball Earth discussed in the previous chapter, it is estimated that at the moment ice coverage reached 30° from the equator, a process of amplification was triggered in which the ever-diminishing solar light absorbed led to a rapid glacial covering of the entire planet, causing the world to be plunged into a wintery realm. Until that time, the situation was more or less reversible. But when that threshold was passed, a domino effect ensued and the climate moved swiftly into a new state of equilibrium. It then took enormous amounts of carbon dioxide to bring the world out of that frozen state.

Sorcerer's Apprentices

So, the greenhouse gas effect is, together with the albedo, one of the crucial elements determining a planet's climate.

It is important to remember that the greenhouse effect is a very good thing for the life on our planet. Without greenhouse gases, the Earth's average surface temperature would be well below 0 °C. Instead, thanks to that ethereal blanket of carbon dioxide, the Earth is a place with moderate temperatures hospitable to life almost everywhere. However, as with all good things, the greenhouse effect must also be used in moderation. Too little and we end up covered in perennial ice. Too much and we risk meeting the same infernal end as Venus.

From here, we can begin to see how playing with the albedo or the concentration of carbon dioxide can lead a species of sorcerer's apprentices to unleash unexpected global consequences that are not easy to manage. A bit like Mickey Mouse with his buckets and mops in *Fantasia*. But, unfortunately for us, there is no wise magician who can come to our aid and put everything back in its place. It is up to us to act wisely.

4

The Explosion that Changed the World

A million years is a long time on the scale of human time. But it is a relatively brief period in the geological history of our planet, which has lasted more than 4 billion years. And yet, in the 100 million years that followed the end of the Snowball, life on Earth flourished, exploring through evolution an enormous number of possibilities and bodily forms, some of which quickly became extinct, while others lasted up to our times, or, rather, became us.

The arrival of these new organisms was influenced by climate and environmental conditions and, in turn, facilitated global changes that would last over time. The close bonds that unite the living and non-living worlds are responsible for the co-evolution of climate and biosphere, and are at the centre of the scientific challenge aimed at understanding how the Earth System functions in its incredible complexity.

From Bacteria to Living Sacks

For more than 3 billion years, life on Earth was predominantly composed of single-celled organisms, with the vast majority of life forms made up by prokaryotes, living beings lacking a distinct cellular nucleus, such as bacteria and Archaea (from which eukaryotes evolved). Bacteria contributed to the great oxygenation event that we discussed in the first chapter, and made the Earth a living planet. That Earth continues to owe many of its characteristics to the prolific and ubiquitous presence of bacteria, from the highest layers of the atmosphere right down to the rocks in the Earth's crust, kilometres below the surface. Our planet is like this because it is populated by innumerable armies of bacteria, as US biologist Paul Falkowski explains. The 'others', from sequoias to jellyfish and elephants, are often kept going by this army of bacteria and would not be able to survive without the incessant activity of the prokaryotic hordes.

In a moment yet to be pinpointed, perhaps around 1 billion years ago, the eukaryotes appeared, organisms with a nucleus that is separate from the rest of the cell. Some eukaryotes began to give rise to compound beings with more than one cell, simple multi-cellular organisms such as the sponges that continue to live in marine waters. After some time, multi-cellular organisms divided into two main kinds: porifera (sponges), and another combination that in turn gave rise to cnidarians (*Cnidaria*) and organisms with bilateral symmetry (*Bilateria*). Cnidarians are essentially small living sacks with an opening that acts as a mouth and an anus, surrounded by tentacles like the hydras that live in fresh water fixed to the floor or submerged objects, or the jellyfish that roam the seas. *Bilateria*, on the other hand, are most other animals, from prawns to horses. In the meantime, the algae also began to develop alternative forms, from red algae to green algae, which dominate in today's seas and fresh water.

Fossil evidence of multi-cellular organisms from before the end of the Snowball was extremely scarce and it was only with the help of molecular biology that multi-cellular beings were discovered to be present before the last global glaciation. This lack of proof could be due to the fact that jellyfish and hydras are less likely to leave fossilized traces. In general, organisms with skeletons, shells or simply more solid parts are much more easily preserved, whereas it is common to see a beached jellyfish disappear in the sun over just a few hours.

Another alternative is that there were not very many of these multi-cellular organisms. It is one thing to be able to evolve into new forms; it is quite another for these new forms to be successful in the competition for resources with pre-existing organisms. Perhaps, a billion years ago, resources were not particularly abundant, competition was too fierce and there was not much space for new experiments in evolution. As such, the multi-cellular organisms would not have been able to 'invent' effective new paths, and would have instead been forced to remain in the simplest forms for millions of years. Perhaps.

Australian Enigmas

Then came the Snowball and everything changed. After the ice melted, a new type of organism slowly established itself, following an initial evolutionary explosion characterized by an incredible variety of new forms, new structures and new kinds of multi-cellular organisms. Perhaps they were (very distant) relatives of our jellyfish. Some are called vendo-bionts (or vendozoans) and belong to the fascinating and enigmatic Ediacaran Fauna, which takes its name from the Australian locality where these fossils were discovered in 1946. Once more, Australia, a continent that has barely been altered by the tectonic processes and geological deformations that characterize most other continents, reveals itself to be a treasure chest of information regarding the Earth's past. Soon after, similar fossils were found in other regions of the world, such as England, close to the White Sea in Russia, China and the Canadian island of Terranova. In addition to this, some fossils discovered in Namibia in 1933, initially interpreted differently, were later recognized as being similar to those found in Ediacara.

The oldest fossils of the Ediacaran Fauna date back to a period between 600 and 540 million years ago (so a few tens of millions of years after the end of the Snowball). And they are strange, very strange. Some are so unusual that it is not even clear whether they were animals, distant relatives of today's cnidarians, or whether they belong to their own kingdom, like fungi, which are neither plants nor animals.

The first interpretation, from Austro-Hungarian paleontologist Martin Glaessner, who later emigrated to Australia, was that these fossils are the most ancient examples of today's animals, the common roots from which everything would later develop. Shortly after, another paleontologist, the German Adolf Seilacher, demonstrated that the Ediacara Fauna was composed of organisms that were actually very distant from the animals that would later evolve. But, in any case, they probably still belonged to the animal kingdom, as a biochemical study by the Australian group led by Ilya Bobrovskiy demonstrated in 2018. The conclusion that they were animals is supported by the presence of chemical traces of steroids (the family of fats such as cholesterol that are typical of the animal kingdom) in *Dickinsonia* fossils, one of the most studied 'Ediacarians'.

The *Dickinsonia* was very different from the organisms that surround us today. It had an extremely flat body and the fossils found to date reveal dimensions ranging from a few millimetres to over a metre. There is no evidence of a structured digestive apparatus or internal circulatory system, and it is believed that this organism was constituted primarily by a membrane that contained liquid at a higher pressure than the water around it, allowing it to move around on the seabed. There is no evidence of a front or back, there is no head, and there are no signs of sensory organs. Perhaps the *Dickinsonia* had a digestive cavity like hydras; maybe it absorbed its food directly from the outside by osmosis through its surface; or maybe it was in symbiosis with the algae that covered its body, providing nutrients through photosynthesis, a sort of autotrophic 'association' between animal and algae. In short, it was a rather alien-like being of which we still know very little.

Other fossils of the Ediacara are even more difficult to interpret. Some had radial symmetry (like jellyfish), others were probably gigantic single-celled eukaryotes (sometimes referred to as protists) with dimensions of some tens of centimetres (protists still exist today, and can reach up to several centimetres). Other were more similar to the animals that came later, like the *Spriggina*, with a bilateral symmetry vaguely similar to that of a mollusc, or the *Cloudina*, a small organism that lived in a tubular shell. So, this was an ecosystem that was entirely different from the one we have today, one shared by organisms that would then disappear and primitive forms of beings that would come to dominate in the future. Indeed, it is so different that some authors maintain that there was not a real Ediacara Fauna, but rather a mix of forms, organisms and attempts by evolution to open new paths.

Avalon

The great explosion of the Ediacara Fauna happened around 575 million years ago, and is known as the Avalon explosion, named after the part of Terranova where the most significant fossils were discovered. Before that period, there is evidence of what would become the Ediacara Fauna, but it is from the Avalon explosion that the wealth and diversity of organisms that populated primitive seas would come. Over a few million years, evolution explored new forms, new structures, new possibilities. Why did this rapid evolution happen? Why, after millions of years of relative evolutionary calm, did all these new life forms suddenly develop? Finding an answer is crucial if we are to understand the workings of the Earth System, the collection of the living and non-living that characterizes our planet.

One possible answer comes from the Snowball itself. In normal times, with a stable climate, natural environments are populated by organisms of every kind that have evolved to utilize the available resources as best they can. There are many feeders and relatively few resources, and the chance mutations that generate new forms of life and new evolutionary solutions have to reckon with an environment that is already crowded. To put it in ecological terms, the 'niches' available to occupy are scarce, making it difficult for the newly evolved organisms to thrive.

During the Snowball, however, many environments became incompatible with life and a great many organisms disappeared. With the melting of the ice, those same environments became habitable once more and ready to host new forms of life that would substitute those that had been lost, and the chance game of evolution was deployed at maximum power. Furthermore, the melting of the ice freed great quantities of sediments and nutrients, the temperatures rose rapidly and the world was once again rich with resources that could be utilized. This is what US geobiologist Andrew Knoll calls 'permissive ecology' – a new world to fill with alternative forms of life. In this context, even those relatively inefficient forms could survive, at least for a bit, and the restrictions on evolutionary solutions were less stringent. When there are lots of nutrient resources and available space, competition is less fierce, and even organisms that do not perform extremely well manage to succeed. In these conditions of abundance, evolution reveals all of its possibilities and natural selection is less drastic.

But there is still one element missing. Even if there were other global ice ages before the last Snowball (the Huronian glaciation, for instance), in the post-glacial periods that followed those episodes there was no particular evolutionary explosion. One explanation, first proposed by Canadian zoologist J. R. Nursall in 1959, is that, before 600 million years ago, the atmospheric oxygen was too scarce to allow for the development of animals of a larger dimension. Today, all the organisms that live in conditions of oxygen scarcity are small or single-celled. It would seem that, in order to allow for the development of a complex animal metabolism like our own (or even just that of the Ediacara Fauna), a true abundance of atmospheric oxygen is required. This only happens in the period immediately preceding the evolution of the organisms preserved in the layers of Ediacara. Permissive ecology and sufficient quantities of oxygen are the two components that we believe are necessary for a thriving development of new life forms.

Very soon (in geological terms), however, even the somewhat alien diversity of the Ediacara Fauna was replaced by a new and even more imposing evolutionary radiation: the era of the Cambrian explosion had arrived.

The Dawn of a New World

According to the geological time scale – one that is undoubtedly conventional but also based on objective characteristics of the stratigraphy of rocks and fossils – the Proterozoic Eon ended around 542 million years ago, making way for the Phanerozoic Eon, which continues today. The first geological period of the Phanerozoic was the Cambrian, which spanned between 542 and 485 million years ago.

At the beginning of the Cambrian, in a few tens of millions of years the majority of organisms from the Ediacara Fauna disappeared, leaving only fossils as a reminder. Later, animals began to develop a plethora of variants, and evolution explored all the possible forms, selecting those that eventually led to us. Burgess's argillites (shale rocks) in Yoho National Park, Canada, are one of the iconic places in which the fossils of the Cambrian explosion are preserved. The Burgess Shale Fauna, discovered in 1909, was only correctly interpreted from the 1970s onwards.

Initially, curious animal forms developed that were still very different from those we see today, though less alien than the previous vendobionts. Like the *Opabinia* with its five eyes, its prehensile trunk and absence of limbs (see figure 4.1). Or the *Anomalocaris* (literally 'anomalous prawn'), a predator some tens of centimetres long with two 'legs' on the sides of its mouth and no other limb except for fan-shaped lobes probably used for swimming. Or the *Hallucigenia* – the name says it all – though for a long time we were unable to decipher even which side was its 'top' and which its 'bottom', which the 'front' and which the 'back', or how it even moved. All of these 'strange marvels', described by great US evolutionary biologist Stephen Jay Gould in his superb book *Wonderful Life: The Burgess Shale and the Nature of History*, came about over a period of 30 million years, starting from the beginning of the Cambrian. A truly explosive period.

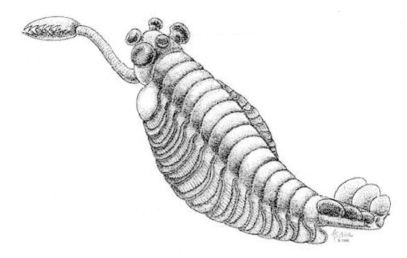

Figure 4.1 A reconstruction of the *Opabinia regalis*, discovered among the fossils of the Burgess Fauna. The *Opabinia*, with five eyes and a prehensile trunk, is the animal most symbolic of the Cambrian explosion and one of the protagonists of the book *Wonderful Life* by Stephen Jay Gould. The drawing is by Marianne Collins.

During that time, all the animal structures that would become widespread in the hundreds of millions of years to come were formed, and evolved into the animals that are now all around us. Some forms, such as the trilobites, splendid arthropods with 3 lobes (hence the name) that became symbolic of the most ancient fossils, survived for around 270 million years before dying out around 250 million years ago. Others modified themselves and became insects, dinosaurs, dolphins or human beings. In that (geologically) brief period at the beginning of the Cambrian, a new world was forged, filled with previously unthinkable possibilities.

Between Avalon and Burgess

An obvious question is why the thriving evolutionary development of the Ediacara Fauna (the Avalon explosion) was interrupted and those living forms, so different from the animals that would follow, disappeared from the face of the planet. There are at least two possible reasons for this, one 'internal' and one 'external' to the biological world.

The internal explanation hypothesizes that the new organisms that were developing, and that would have become enormously widespread during the Cambrian explosion, were superior (in competitive terms) to Ediacarian living forms. When an organism that knows how best to exploit resources eliminates the other organisms that are competitively less brilliant, we are faced with a substitution that is entirely internal to the mechanisms by which ecosystems and evolution function. This perspective, however, does not take into account the fact that the Cambrian explosion does not happen simultaneously with the disappearance of most of the Ediacara Fauna. Furthermore, the development of the new Cambrian organisms seems to be due to the opportunity to occupy newly empty niches, rather than to the competitive substitution of less well-performing organisms.

It is here that the second possibility comes into play: an external catastrophic event, brief but very intense, that could have swept away a large part of the Ediacarians, leaving a large number of ecological niches free once more, ready to be filled with the product of a new, random evolutionary explosion. Indications from the isotopic analysis of rocks that formed just before the Cambrian suggest that the oceans had become anoxic, extremely poor in oxygen. We do not know what caused this condition. Perhaps a rapid and unexpected emission of methane from the sediments, or maybe a global ocean mixing event that brought deep waters that are low in oxygen to the surface. Or maybe the impact of an asteroid or comet.

In any case, the message is that, each time new possibilities are freed up, evolution is able to rapidly fill the empty space that has been created, showing great 'imagination'. Every catastrophe is followed by the development of a renewed world different from the one that has gone before it. This happened with the Snowball; it happened between the organisms of Avalon and Burgess. In the next chapter, we will see other examples of catastrophes and the ensuing new possibilities. Even if we were to continue doing our best to ruin our planet's environment, exterminating those that travel alongside us, reducing biodiversity and devastating the climate, it would cause only temporary damage. Within a few million years, evolution would take a different path, recreating new environments and new organisms. But we will certainly not be around to see it.

Punctuated Equilibria

In 1972, paleontologist Niles Eldredge and biologist Stephen Jay Gould, whom we have already cited, published an important work in which they discussed how evolution did not progress gradually but was dominated by intermittent episodes (events that 'punctuate' the course of geological history). During these relatively brief periods, new forms of life developed rapidly, along with new body structures and new physiological and biochemical solutions.

At the beginning of geological history, the vision of 'catastrophism' prevailed. It was believed that the great geological structures had been generated by large-scale catastrophic events, such as biblical floods, landslides of colossal proportions, earthquakes, enormous volcanic eruptions. At one time, it was believed that the Earth was 'young', only a few thousands of years old. Some, imaginatively interpreting Bible passages, even believed they were able to determine the exact day on which the planet was formed. For example, in 1650, the Irish Anglican reverend James Ussher fixed the date of the Earth's creation as 23 October 4004 BCE. Others argued over slightly different dates, but always dating to just a few thousands of years ago. They simply could not understand how 'normal' forces could have generated such impressive structures.

It was Charles Lyell, the Scottish scientist who founded modern geology, who, halfway through the nineteenth century, developed the concept of gradualism in Earth sciences, extending and completing the ideas that had already been expressed by geologist James Hutton. Essentially, the concept is that the great geological structures, such as mountain ranges, were not generated by exceptional causes but by 'normal' forces that acted over an extremely long period of time. Not thousands of years, but millions or hundreds of millions of years. The time scale had suddenly expanded, and the abyss, which was in some senses frightening, opened up by the extent of geological time left those who were used to thinking in terms of centuries, or at most millennia, feeling both disconcerted and full of admiration. Still today, if we take our personal experience as a basis, we have great difficulty understanding what 100 million or a billion years really means. Human time and that of the planet pass in a different way, and it is not always easy to perceive and experience both harmoniously.

Lyell's work inspired that of Charles Darwin, with his theory of evolution taking evolutionary gradualism as its conceptual cornerstone. This is the slow, random unravelling of new possibilities while natural selection works at pruning, eliminating those variants less suited to survival. It is clear that, in this context, the article by Eldredge and Gould created turmoil, but it was a turmoil that brought new ideas. Geological, and biological, history is truly very long, and changes happen continuously and gradually. But superimposed on this are moments from the Earth's history during which exceptional events, both internal and external, led to an extreme acceleration, rapid change, impetuous and unexpected developments. Later on, we will see how the climate dynamics also alternate between moments of gradual, slow and generally predictable variation, and situations of intense and fast change, often generated by external perturbation or by having reached a critical threshold in some climate processes. We have seen this with the triggering of the Snowball and with the rapid deglaciation of 635 million years ago, and we will find it again in geological periods closer to our own.

In the philosophy of science, a similar position was held by Thomas Kuhn, who identified 'normal' science – times in which development is gradual and previously developed ideas are refined – and moments of scientific revolution, in which old, now obsolete ideas are swiftly abandoned and new paradigms developed and submitted to factual verification. Preservation and revolution, the two elements that regulate the history of the world like yin and yang.

5

Between Catastrophes and Opportunities

With its emergence from the Snowball and the Cambrian explosion, the planet had taken on a new form. Evolution was unfurling its wings and a plethora of bodily structures and life forms were appearing on Earth.

Many things happened in the 500 million years that followed. The movement of the continents, shifted by the convective motions in the mantle, generated new lands and new mountain ranges that were then eroded and fragmented. The climate continued its oscillations between hot and cold, and life fully conquered the continental lands that emerged from the oceans.

During all of this, a number of large-scale catastrophes caused mass extinctions and were then followed by new evolutionary impulses that repopulated the environment with increasingly diverse and unexpected organisms, thanks to the random nature of mutations and the needs of natural selection.

The interaction between the living and the non-living played an even more vital role, and carbon dioxide continued to be one of the great drivers of the planet's climate.

The Climate during the Phanerozoic

In the timeframe dating from the beginning of the Cambrian (around 540 million years ago) to the beginning of the Eocene (around 55 million years ago), the climate of the Earth was, on average, hotter than it is today, with very little ice presence. There were also some cold episodes with extensive glacial coverage, but no more global glaciations like during the Snowball. The continents were no longer amassed around the equator, atmospheric methane had by then been reduced to minimal levels and oxygen continued to increase.

During this period, the concentration of carbon dioxide in the atmosphere was the main factor controlling the climate (though not the only one, of course). Variations in the quantity of carbon dioxide were determined by fluctuations in the carbon cycle, in turn associated with temporary (in a geological sense) imbalances such as volcanic emissions into the atmosphere, the removal of carbon dioxide through the weathering of superficial silicate rocks, and the burial of organic matter and carbonates in the seabed. The greater or lesser presence of broad areas of continental shelf, the tectonic processes involving the aggregation of the plates into supercontinents or the fragmentation of the continental masses, the formation or erosion of the mountain ranges all contributed to oscillations in the amount of carbon dioxide present in the atmosphere and, as a result, modified the greenhouse effect, pushing the climate alternately between cold periods and warmer eras with no ice. Before 50 million years ago, there were at least three periods associated with extensive glaciation: around 450 million years ago, 300 million years ago and 150 million years ago.

These glacial episodes could have been triggered by low-level carbon dioxide emissions by volcanoes, or by its speedier removal through alteration of the silicates. The most traditional interpretation, already expressed by US geologist Thomas Chamberlin in 1899, is that these glaciations were unleashed by a faster surface weathering of the rocks, as we have already seen with the Snowball. Many other researchers have confirmed this interpretation, explaining the acceleration in the weathering of the rocks with the 'temporary' passage of some of the continental masses across equatorial zones, where the temperatures are higher and precipitation more intense. However, in 2016, Ryan McKenzie and his team at the University of Hong Kong published an article showing how the glaciations of the Phanerozoic were also associated with lower levels of volcanic activity in the periods immediately preceding them, leading to a reduction in the levels of carbon dioxide being released into the atmosphere. Continuing the saga, in 2019, other scientists studied in great detail a number of ophiolites (rocks with a silica content lower than around 50 per cent, which come from the oceanic crust and upper mantle) that had been exposed on the surface at length, and demonstrated that the three glaciations cited before, and the fourth – which began around 35 million years ago and is still in progress – were preceded by the shift of continental masses across the equatorial belt, giving credence to the hypothesis that the climate is (also) controlled by the geochemical removal of atmospheric carbon dioxide. Or, rather, biogeo-chemical removal, given also the actions of the plants that had already started to colonize the continents.

These climate processes are extremely complex, and only now are we beginning to explore and understand them. It is therefore entirely natural that the hypotheses overlap and that new evidence shifts the balance in favour of one explanation or another. It is also possible that there is more than one cause, and that, as is common in complex systems, different contributing causes act simultaneously. In any case, there are two important lessons that we have learned from this research: that the concentration of carbon dioxide in the atmosphere is a crucial element for our planet's climate (at least since 600 million years ago when methane became less relevant), and that the geology of our planet is inextricably linked to the climate, the atmosphere and life on Earth, creating a single, living planet.

Measuring Geological Time

Before going any further, it is worth pausing and asking ourselves how it is that researchers are able to obtain information on the climate of the past. Similarly, how can we know that those rocks which hold a wealth of incredibly useful information on the climate have a particular age? There are many methods and one of the most effective utilizes isotopes.

Matter is made up of atoms, which contain a nucleus composed of protons and neutrons, and an external shell formed by an electron cloud. Take oxygen, for example. Here, the electronic cloud has a diameter of around 1 ångström, which is about 100 billionth of a metre (or 1 metre divided by 10 billion, $1/10^{10}$). The atom's nucleus is, instead, much smaller, around 100,000 times smaller than the dimensions of the atom.

The protons carry a positive electric charge, while electrons have a negative charge that is equal and opposite to that of the protons. A neutral atom has an equal number of protons and electrons. In the nucleus, the positive charges of the protons generate a repulsive force that sharply pushes them away from one another. However, the presence of neutrons, which have no electric charge, holds the protons together thanks to the strong nuclear force, which is more intense than the electromagnetic force at the very small distances within the nucleus's dimensions. In order to obtain a stable configuration in those nuclei composed of more than 1 proton (all elements except hydrogen), the number of neutrons must generally be at least equal to the number of protons (with the sole exception of helium-3, whose nucleus is composed of 2 protons and just 1 neutron). The greater the number of protons in the nucleus, the greater the number of neutrons required to balance the repulsive effect of the electromagnetic forces.

The properties of the various chemical elements, from hydrogen to plutonium, are determined by the number of protons and the number of electrons. For example, the nucleus of lithium has 3 protons, carbon has 6 protons, while oxygen has 8. The number of protons in the nucleus determines the 'atomic number' of every element, while the total number of protons and neutrons determines its 'mass number'.

Many elements (though not all) can exist in forms with a different number of neutrons, known as isotopes. For example, the nucleus of hydrogen can have 1 solitary proton, or 1 proton and 1 neutron (as is the case with deuterium), or just 1 proton and 2 neutrons (tritium). Figure 5.1 illustrates the isotopes of hydrogen.

The helium most commonly found on our planet has 2 protons and 2 neutrons (helium-4, ^{4}He), but there is also a helium with 2 protons and just one neutron (helium-3, ^{3}He). Similarly, oxygen can exist with 8

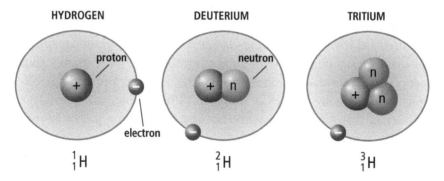

Figure 5.1 Schematic illustration of hydrogen isotopes (deuterium, ^{2}H, and tritium, ^{3}H). The nucleus and the electron cloud are not to scale. In reality, the electron is much farther from the nucleus than shown here. *Source:* www.chimica-online.it/download/isotopi-idrogeno.htm.

protons and 8 neutrons (oxygen-16, ^{16}O), or with 9 neutrons (oxygen-17, ^{17}O) or even with 10 (oxygen-18, ^{18}O). The small number in the top left of the element's symbol indicates its mass number. To add another element that is crucial to our research, carbon can exist in forms with 6 protons and 6 neutrons (^{12}C), or with 7 neutrons (^{13}C) or with 8 (^{14}C).

The isotopes of a particular element have substantially the same chemical properties, but they differ in mass. Some, such as tritium (^3H) and carbon-14 (^{14}C), are unstable. This means that, over a varying timeframe, depending on the type of isotope, they decay into other elements. They are, therefore, radioactive isotopes. For example, in the decay of carbon-14, a neutron is transformed into a proton, producing a nucleus of nitrogen-14 (with an atomic number of 7 and a mass number of 14), an electron and an antineutrino (a neutral elementary particle with an extremely small mass), and releasing energy.

The decay of a radioactive element is a process regulated by the laws of probability. If we have a single radioactive nucleus, we do not know when decay will occur. But we can precisely determine the average time it takes for the nucleus to decay. If we put lots of radioactive nuclei together, some will decay earlier, some later, but on average we will be able to say when that initial number will reduce by half. And then by half again. The 'half-life' is the time in which the mass of the radioactive material reduces by half, and it is independent of any external chemical or physical conditions.

Often, there is not a single decay of a radioactive isotope, but rather a sequence of successive decays until a stable isotope is eventually produced, which we can call the 'daughter' (or, rather, granddaughter). For example, uranium-238 (parent) goes through a series of spontaneous decays that eventually produce the stable isotope lead-206 (grand-daughter). Knowing the decay chain and the half-life, comparing the quantity of daughter isotope with that of the remaining parent isotope allows us to determine the sample's age. Here, we are entering into the world of radioisotope geochronology, based on accurately establishing the chemical and isotopic compositions of minerals and rocks. It is one of the sciences on which geological dating is based and, therefore, the starting point for reconstructing the events of the past.

It is not all easy, of course. For example, we cannot always use the same radioactive elements. If the decay time is very brief in comparison to the age of the rock, very little parent radioactive material will remain and we will make a significant error in our estimation of the time passed. Equally, if the decay is too slow, too little of the daughter isotope will be formed and we will once again have significant errors in our time estimation.

For this reason, in order to date relatively recent materials, like prehistoric or historic matter, carbon-14 is used with its half-life of 5,730 years. However, in order to determine the age of much older rocks, we might use potassium-40 with its half-life of more than a billion years, or samarium-147, which has a half-life of around 106 billion years. These are often used to analyse zircons, extremely resistant and long-lasting crystals that hold the isotopic information that was present at their formation. Contrary to what was stated in an old advertisement for diamonds, it would be better to say 'A zircon is forever.' Each age has to be assigned its own appropriate material, but today's researchers are nevertheless capable of dating the Earth's past with great precision.

Stable Isotopes

The age of a sample is essential, but it does not directly provide information about the climate at that time. In order to obtain this, we must instead employ other methods based on stable isotopes, those that do not decay spontaneously. For example, the three stable isotopes of oxygen – oxygen-18 in particular – provide us with a key for understanding both temperature fluctuations and continental ice cover. Modern geochemistry using stable isotopes and its applications for the reconstruction of oceanic temperatures began in the 1940s with the work of Harold Urey at the University of Chicago, whom we have already encountered looking at laboratory experiments on the origins of life. Researcher Cesare Emiliani, who was born in Italy before emigrating to the US for work, soon became a pioneer in the use of isotopes for paleoclimatic research, consolidating the basis of a science that today is crucial for understanding the variability in the Earth's climate.

In sea water, there is predominantly oxygen-16, which is the most abundant form. There is also oxygen-18 in lower quantities. As oxygen-18 is heavier, during evaporation those water molecules containing oxygen-16 are favoured as they are lighter and therefore more volatile at the same temperature. The temperature is linked to the energy of the molecules' thermal agitation, which in turn influences the probability that a molecule might be able to escape the bonds of the liquid phase during evaporation. We therefore have an example of 'isotopic fractionation': the evaporation process that favours the presence in water vapour of one isotope (^{16}O) over another (^{18}O). Water in the clouds becomes richer in oxygen-16 than the sea water. If the evaporated water and the successive precipitation remains trapped in polar ice sheets and in mountain ice, then the excess of oxygen-16 in precipitation does not return entirely to the sea and, as a consequence, the concentration of oxygen-18 in marine waters is higher than we would see in warmer conditions. Therefore, the isotopic ratio between oxygen-18 and oxygen-16 in the marine waters of a particular place, or during a particular period, provides us with infor-mation on the climatic conditions at that time.

The isotopic ratio between oxygen-18 and oxygen-16 is also maintained in the organisms that live in the sea, especially in the shells some of them build – in particular, foraminifera, single-celled protozoa that make up the zooplankton or live on the seabed and form carbonate casings that maintain the ratio between these two oxygen isotopes. As we have already seen when discussing the geological carbon cycle, when foraminifera die, their casings are deposited along the seabed and contribute to the formation of sedimentary rocks, often remaining recognizable as micro-fossils and aiding the process that removes carbon dioxide from the atmosphere. When we drill the seabed and extract a 'core' from these rocks (typically, a long vertical cylinder obtained through drilling), the ratio between the two oxygen isotopes obtained from the microfossils provides us with an estimate of the climate conditions at the time of the sediment's deposition, which can be dated using the geochronological methods described earlier.

Different isotopes provide a broad spectrum of information on the climate and the environment. Without losing ourselves in the meanders of isotopic geochemistry, let's take a look at another important example, one that involves carbon. The two stable isotopes of carbon are carbon-12, the most abundant, and the rarer carbon-13, which has an extra neutron and is therefore heavier. Photosynthesis favours the absorption of carbon dioxide molecules containing the lighter carbon-12. The herbivores that eat the plants, and the predators that eat the herbivores, maintain the same isotopic ratio of the two types of carbon. The ratio between the abundance of the two isotopes is therefore different in biological matter and in the rocks generated by inorganic processes such as those that take place in the mantle. If part of the organic matter produced by photosynthesis is buried, in periods of great biological productivity a great deal of carbon-12 is absorbed and held on to by organic matter, only returning to circulation later. The inorganic sediments deposited in those periods will therefore have an isotopic ratio richer in carbon-13 than occurs in periods with scarce productivity. However, large-scale volcanic events return the ratio between carbon-13 and carbon-12 in the atmosphere to levels closer to those found in the rocks in the mantle. Finally, a net reduction in the isotopic ratio of carbon-13 to carbon-12 indicates a sizeable emission of carbon dioxide or methane from organic sediments, as happens today due to the use of fossil fuels, and as happened in the great extinction at the end of the geological period known as the Permian, which we will look at shortly. By determining the relative abundance of different carbon isotopes, we can obtain relevant information on how active photosynthesis was, how much greenhouse gas was released by decaying organic matter, the possible eruptions that took place in the past, and the periods of biological crisis.

Land Ho!

In the seas more than 500 million years ago, the Cambrian explosion gave rise to great biodiversity featuring articulated and unusual (for us) forms and structures, as we saw in the previous chapter. However, it was essentially only bacteria and photosynthetic cyanobacteria that lived on land at that time. Even the soil was different. It wasn't the rich heterogeneous compound formed by living creatures and filled with organic substances that we see today, but poor soil dominated by mineral components, with bacteria that formed prokaryote ecosystems in the interstices of sands, swamps and lakes, and perhaps lichens on rocks. This soil was probably rather similar to what we find in deserts today, with the sand compacted on the surface by a bacterial crust.

So when did the first eukaryotes appear on land? In order to live outside of the water, the organisms had to adapt to difficult conditions. The Sun's ultraviolet rays had a sterilizing effect, nutrients were scarce and the risk of desiccation for cells was high. In 2016, British paleontologist Martin Smith described what was believed at the time to be the oldest fossil of a complex land-dwelling organism: a small fungus called *Tortotubus*, dating from around 440 million years ago. The eukaryotes had started to conquer the continents, thanks also to the rise in the concentration of oxygen in the atmosphere and the subsequent growth of the protective ozone shield that safeguards the surface from ultraviolet rays. But, as always happens with scientific research, new discoveries can change how we see the world. In 2019, Belgian paleobiologist Corentin Loron and his colleagues discovered traces of an even older land fungus that lived almost a billion years ago in the Canadian Arctic (that's even before the Snowball!). And in 2020, the research group of another Belgian, Steeve Bonneville, discovered evidence of a fungus that lived around 800 million years ago in what is now Congo. So, the challenge is ongoing and all of these new observations must be carefully evaluated and verified before drawing any definitive conclusions.

In any case, it is now clear that fungi were the first complex organisms to occupy the continental regions, preparing the terrain (literally!) for what would follow. The eukaryotes rapidly (in evolutionary terms) occupied terrestrial niches, diversifying, expanding and making the most of the opportunities they were offered by the new environments there to be colonized. Around 400 million years ago, gigantic mushrooms known as *Prototaxites* dominated the landscape. These cylindrical mushrooms measuring a metre in diameter and up to 9 metres in height were discovered midway through the nineteenth century. These structures were composed of numerous fine, interlacing filaments and it is believed that they were characterized by a symbiosis between the fungi and single-celled algae capable of photosynthesis. If this is the case, we should instead consider them lichens, albeit seriously super-sized relatives of the pioneer lichens we see today on rocks in the mountains.

In that same period, plants evolving from algae began to colonize the land. Once again, discoveries of the 'oldest plant' followed in quick succession, calling on the scientific community to verify the results and ultimately review the dates for the history of life on Earth. To date, the oldest known plant fossil is the *Cooksonia*, named in honour of the Australian paleobotanist who studied them in great detail: Isabel Cookson. These vascular plants (meaning they have an internal system for the circulation of water) appeared at least 420 million years ago and were only a few centimetres tall. They had no leaves, no flowers, no seeds, and some species of *Cooksonia* perhaps coexisted in some environments with the towering *Prototaxites*. In short, this landscape looked like something out of a fantasy novel.

In reality, some fossil spores of non-vascular plants (such as the more primitive mosses) date back to around 470 million years ago, suggesting that plants had begun to colonize the continents even earlier. But why, in a book on the climate, are we interested in knowing all of this? It is because the arrival of plants on the continents likely played a crucial role for the climate. It changed the albedo, it changed the water cycle and the speed at which rocks were weathered. The plants absorbed carbon dioxide from the atmosphere, storing it in their fibres and then in the subsoil. Once again, living organisms were the engines driving change in the climate.

Mosses, Forests and Ice

As we discussed at the beginning of the chapter, around 450 million years ago, the Earth's climate cooled once more, ushering in the first glaciation of the Phanerozoic. In the previous period, the concentration of carbon dioxide in the atmosphere was at least 10 to 15 times higher than it is currently, reaching more than 4,000 ppm. In order to trigger a glaciation, it was necessary for CO_2 concentration to drop to no more than 8 times the current amount. But the geological mechanisms based on the increased weathering of the rocks by the movement of continents in equatorial zones, and/or lower volcanic emissions, do not seem to have been sufficient to cause such a drastic drop in CO_2 concentration.

In 2012, lecturer in climate change at the University of Exeter Tim Lenton, together with a substantial team of colleagues, suggested that the arrival of non-vascular plants (such as mosses) on land was what triggered the glaciation. Indeed, mosses accelerate the weathering of silicate rocks, with the consequent absorption of carbon dioxide from the atmosphere. Experiments carried out in the controlled confines of a laboratory, where it was possible to measure precisely the amplification effect caused by the moss's presence, confirmed a marked difference between rock alteration phenomena associated with plants and those that are purely inorganic. If this interpretation is correct, then the first glaciation of the Phanerozoic could have also been triggered by the arrival of the first plants on the continents.

A similar mechanism was at work in the second 'modern' glacial episode around 300 million years ago. Already in 1997, US geochemist Robert Berner had proved how the development of forests filled with vascular plants and active soils further accelerated the weathering processes, even more so than the moss. And, without a corresponding rise in volcanic emissions, the absorption of carbon dioxide by the plants led once again to a glacial era that, like the previous one, was not global but rather similar to the oscillations seen in our time, which we will discuss shortly. The mechanisms at work are undoubtedly many, but, nevertheless, the evidence for living organisms having an effect on the climate is increasingly persuasive.

Things Do Not Always Go to Plan

If, as we have seen, biological processes can change the planetary environment and the climate, it is equally clear that certain geological or astronomical events can have an extremely powerful impact on life and ecosystems. Little more than 250 million years ago, there was a colossal mass extinction event in which up to 90 per cent of species, both marine and terrestrial, were wiped from the face of the Earth. This happened between the end of the geological period we now call the Permian and the period known as the Triassic. Dozens of species of trees, which had evolved to form thick forests, and small predators such as the *Dinogorgon* – a therapsid not too evolutionarily distant from mammals – were swept away by a catastrophe worthy of the Hollywood treatment. But what happened? How did all this come to pass?

It is much like a cold case in a police drama: who or what almost eliminated life from Earth? Researchers' first thought was that it was an asteroid impact, or a sudden release of methane from the seabed. The sediments deposited during the extinction are characterized by a low isotopic ratio between carbon-13 and carbon-12, indicating the presence of volcanic eruptions or a release of carbon trapped in organic matter such as coal. Furthermore, there is evidence of a high concentration of mercury in many areas across the planet. But it is volcanic eruptions that are held today to be the main suspect. Around 251 million years ago in Siberia, there was a series of gigantic eruptions that left impressive geological evidence in what we know today as the Siberian Traps (the geological term *traps* comes from Swedish and means 'steps', referring to the great stair-like structure of this landscape). For almost 2 million years, basalt lava covered the Siberian terrain and super-volcanoes threw enormous quantities of dust, ash and carbon dioxide into the atmosphere. The climate changed suddenly and the rain became acidic, devastating forests and plants and upsetting all of the ecosystems and trophic networks based on primary production by vegetation. It is likely that vast deposits of coal burned for thousands or hundreds of thousands of years, releasing huge quantities of carbon into the atmosphere. The world was, once again, almost finished.

Some, however, gained an advantage from this extermination. Fungi, for example, found themselves dining on dead wood for millennia, aiding the decomposition process and leaving us an enormous quantity of fossil evidence of their considerable abundance. And once again, over time, the ecological niches left free by the lost species became the theatre for new evolutionary processes, new organisms and a new environment that would lead, over the millions of years to come, to that world of dinosaurs that appeals so very much to our imaginations.

The first dinosaurs appeared around 240 million years ago in the Triassic period, occupying evolutionary niches that the great extinction had freed up. For tens of millions of years, the dinosaurs dominated (in a media-friendly sense, at least) the lands and the seas. Some dinosaurs had feathers, and many were almost certainly warm-blooded and looked after their young. Up until around 100 million years ago, the climate was for the most part temperate and cool, moving then towards a hot (and later, even hotter) period that began halfway through the Cretaceous period.

But the world of the dinosaurs also met with a brutal demise. Around 66 million years ago, an asteroid with a diameter of around 10 kilometres hit the Earth, triggering a series of consequences that would end the dinosaurs' reign. Once again, an external event profoundly modified our planet's environment. In 1980, US physicist Luis Álvarez (awarded the Nobel Prize in 1968) and his geologist son Walter described the presence of a layer of clay rocks rich in iridium (a very rare element on the Earth's surface but abundant in asteroids) dated as 66 million years old, indicating a time at the end of the Cretaceous and the beginning of the Paleogene. This layer can be seen in various locations all over the world suggesting the event was a global one, and it was first observed at the Gola del Bottaccione near Gubbio, Italy, a place of incredible geological value, both scientific and emotional, that can still be easily visited today. This layer, shown in figure 5.2, is important because it suggests that it was the impact of an asteroid that exterminated the non-avian dinosaurs, while the avian dinosaurs followed an evolutionary path that led to the birds we know today. Many other life forms disappeared along with the dinosaurs, such as the ammonites that had characterized the seas of the Cretaceous.

Figure 5.2 The layers of the Gola del Bottaccione, near Gubbio, where Álvarez father and son first observed the layer of clay containing iridium, suggesting the likely impact of an asteroid and the subsequent disappearance of non-avian dinosaurs. The boundary between the layers from the Cretaceous and those from the Paleogene is where the lighter rocks end. It is here that we find the layer of clay, around a centimetre thick and rich in iridium. The horizontal dimension of the scene captured here is around a metre and a half. Author's photo.

In the years following the Álvarez discovery, a gigantic impact crater of the same age as the layer of iridium was identified in the Gulf of Mexico. The guilty party was, thus, identified. The impact threw enormous quantities of dust into the atmosphere, darkening the sky for years and leading to the collapse of photosynthesis and the ecosystems. The smallest animals survived the catastrophe, but the dinosaur world had reached its end.

However, as always, things in nature are far more complicated than they might seem at first sight. Around 66 million years ago, there was an enormous volcanic eruption in India that created what we know today as the Deccan Traps. This eruption may also have contributed to the extinction of the dinosaurs, with similar mechanisms to those we saw in action at the end of the Permian. Of course, it is difficult to imagine how unlucky it was to have the impact of a large asteroid and the eruption of super-volcanoes happening so soon after one another, but it is possible. Equally, the incredibly violent impact of the asteroid, which will have undoubtedly generated powerful earthquakes, could have disturbed the equilibrium at the Earth's mantle, unleashing or amplifying the Deccan eruptions.

In any case, the disappearance of the dinosaurs paved the way for other beings, who made the most of the available resources. This is how the ascent of mammals began, already present before the catastrophe but inhibited by the dominion of the great reptiles. The history of humanity, therefore, begins with an asteroid impact (a random event) that paved the evolutionary way to the first hominids and then to us. If this impact had never happened, the planet would perhaps be populated by a species of intelligent reptiles.

From these events, we learn how life on Earth and its environments are exposed to sudden and unpredictable catastrophes, and that until now (luckily for us) the end of a world has never been so absolute as to cancel out the biosphere. Life and evolution always managed to exploit to the best of their abilities the opportunities generated by catastrophes in order to create new environments, different from those that went before. However, the price to pay is the loss of a great number of individuals, species and entire ecosystems.

6

The Living Planet

Over the last two chapters, we have seen how the physical, chemical and geological environment, with its catastrophes both internal and external, forged the biosphere of our planet. In turn, living organisms have influenced many characteristics of both the environment and the climate, from the concentration of oxygen in the atmosphere to the global carbon cycle. While the surface's albedo and the composition of the atmosphere are two of the climate's main drivers, the presence of life undoubtedly modifies, modulates and enriches them. This is the Earth System, in which the atmosphere, the oceans, the geosphere and living organisms perpetually interact.

A Grand Organic Whole

Between the end of the eighteenth century and the middle of the nineteenth, great German naturalist Alexander von Humboldt explored various regions of the world, collecting information, data and knowledge on the functioning of our planet, from the volcanoes to the magnetic field, from living species to their distribution over different climates and altitudes. Von Humboldt was a polyhedric, romantic and adventurous scientist who became one of the most famous men of his time and who, at the end of his life, published a great work over five volumes called *Cosmos*, in which he attempted to lay out everything he had discovered about our planet and the complex bonds between the living and the non-living known at that time. With these books, von Humboldt laid the foundations for what would later become Earth System sciences, which are at the centre of today's research on the climate and the environment.

In his view, nature was one great chain of causes and effects, in which no single event or fact could be isolated from any other. He perceived nature as a global force, following an approach that was revolutionary for those times. The German scientist essentially defined the modern concept of nature as we understand it today. From the interconnectedness of organisms came the notion of the trophic network, with its complexity and the possibility that changes in single components can influence the entire system. Von Humboldt was also one of the first scientists to be concerned about the environmental changes caused by anthropic activity, observing the devastation caused by colonial plantations in Venezuela and reflecting on the danger of climate changes brought about by thoughtless human actions, such as uncontrolled deforestation. He was a giant of science, immersed in his time but capable of speaking to later generations. He is, however, less well known today than he should be, though a recent book by Andrea Wulf offers him the recognition he deserves.

Shortly after, in 1876, Alfred Russel Wallace, the English scientist who, alongside Charles Darwin, was jointly responsible for the theory of evolution by natural selection, published the book *The Geographical Distribution of Animals*. In this volume, Wallace revisited the concept of complex interactions and mutual interdependence that bind plants and animals with each other and the environment in which they live, making our planet 'one grand organic whole'. Once more, we have a sense of the co-evolution of the geosphere and the biosphere, of the environment and living organisms.

Our understanding of the relationships between living organisms and the physical and chemical environment took a step further with the work by the Russian Vladimir Vernadsky, one of the founders of modern biogeochemistry. In Leningrad in 1926, Vernadsky published a fundamental text titled *The Biosphere*, translated into French in 1929 and into English in only 1979. This volume discusses some of the principal concepts developed by Vernadsky. Namely, that life is the principal agent for geological transformation and that the effect of living organisms has grown throughout the geological history of the Earth. The notion of the biosphere, defined by the Austrian geologist Eduard Suess as 'the part of the Earth's surface where life prospers', was taken up and expanded by Vernadsky. In this understanding, the biosphere is not static but grows over time and involves the atmosphere, oceans and rocks, because all of these components, and the chemical reactions they produce, are dominated and controlled by the activity of living organisms.

In reality, the Russian geologist and geochemist pushed further and hypothesized the evolution of the Noosphere, which is the transformation of the biosphere into a new form dominated by human rationality, knowledge and action. Interestingly, this concept, developed by the atheist Vernadsky, resonates strongly with the philosophical ideas of Pierre Teilhard de Chardin, a French paleontologist and Jesuit, a charismatic and complex figure who often met with hostility from both scientists (who rejected his finalistic view of evolution) and the religious hierarchies of the time (who rejected his unorthodox theological views). The notion of the Noosphere seems to act as a prelude to the idea of the *Anthropocene*, the geological era dominated by human action which we will discuss later on, and even that of the evolution of new forms of life that are not necessarily biological but perhaps somehow robotic, as spoken about by English geochemist James Lovelock in his last book, *Novacene*.

But let's go back to the ecosystems and their capacity to influence the climate. The term 'ecosystem' was introduced in 1935 by Arthur Tansley, who described it as 'the whole system (in the sense of physics) including not only the organism-complex, but also the whole complex of physical factors forming what we call the environment of the biome', in perfect harmony with the visions of von Humboldt, Wallace and Vernadsky. So, in order to explore how the biosphere influences the climate, let's start by seeing what happens on a small scale (in a valley or on a hillside in the desert) before then expanding our discussion to take in the entire globe and returning to some of the scientists and concepts we have just discussed.

Ecosystem Engineers

One of the conceptual cornerstones of ecology is that organisms adapt to occupy an 'ecological niche', an environment characterized by a collection of very precise physical, chemical and climate properties. There are plants and animals adapted to life in the equatorial forests, others that live in the desert, others again that prosper in the Arctic and the Antarctic in conditions that are often prohibitive. We find proof of this by moving upwards from the flat of a mountain valley. Both the vegetation and the fauna change, their variations dependent on altitude, the abundance of water, the extent to which the slopes are exposed, the properties of the soil and the kinds of rocks found beneath it.

But the issue is, as ever, more complex. As we have said, the organisms not only adapt to environmental conditions but, when they are able, they modify them in order to guarantee a greater chance of survival and continuation of the species (or, better, of their genetic pool). It is not only, therefore, a passive making the most of a pre-existing environment, but an active and constant effort to modify the world to their own advantage. For us humans, used to changing the environment we live in since our species's arrival, this should be no surprise.

In 1994, this concept was clearly expressed in a paper by ecologists Clive Jones (US), John Lawton (English) and Moshe Shachak (Israeli), which coined the term 'ecosystem engineers' in reference to those organisms that significantly modify the environment. Human beings aside, one clear example of this is beavers. When they arrive in a mountain valley, they begin to build wooden dykes, bringing down trees and creating a lake. In this way, they transform a terrestrial environment into an aquatic one, which is much better suited to them and capable of protecting the beavers from most land-based predators. The overall result is that the modified environment provides a disadvantage for the survival of land-based species, favouring instead the presence of plants and animals that live in the water. By modifying the environment to their advantage, beavers influence the entire ecosystem and the other living organisms which inhabit it, at times excessively. In regions where they have been introduced by humans, beavers can overmodify the environment (building dykes and chopping down trees) and their population must be reduced and controlled, as happens in Argentinian Patagonia. Perhaps they remind us of another group of living beings who tend to modify the environment they live in a little too much.

Desert bushes and their interaction with the bacterial crust provide us with another example. In arid areas such as the Southern Negev, in Israel, with around 100 mm of annual rainfall concentrated in just a few intense events, the undisturbed sand is compacted, forming a physical crust that is soon colonized by cyanobacteria, which transform it into a biological crust. During dry periods, the bacteria are often in a dormant state but ready to reawaken quickly with the first sign of rain. Here and there, the crust is interrupted by the presence of bushes. The bacterial crust reduces the permeability of the surface to the infiltration of rainwater. In this way, the scarce precipitation is partially channelled towards the bushes, which collect much more water than they would have been able to had there not been a crust. Close to the bushes, an island of humidity and fertility is formed, attracting other plants and animals. Isopods in particular, small land-dwelling arthropods, help to move the soil and favour infiltration close to the bushes. Acting in their own best interests, in synergy with the biological crust, the desert bushes alter the environment by creating (at least in part) their own ecological niche and influencing the characteristics of the local micro-climate.

This is where the important concept of the *construction* of an ecological niche by certain species of living organisms arises. The beavers create their own aquatic environment, which is safe and better suited for them, while desert bushes work with the biological crust to create fertile islands. This process also involves a hereditary transmission that is not directly genetic. The new generations that grow in the transformed environment are at an advantage and can have an easier life (as it were) than the first generation of colonizers. The matter is, of course, much more complicated, because an organism's tendency to behave as an ecosystem engineer is always governed by the genetic transmission that continues to drive evolution, but the inheritance of a more appropriate environment undoubtedly helps the young and modifies the selective pressures. It is a little like what happens with humans. Being born into a family with a sizeable economic legacy at their disposal can provide a very different start from that of someone born into poverty. Then the game of life takes its course – often proving unpredictable for both beavers and humans – but the initial conditions are different.

One important point is that beavers and bushes do not 'know' that, by modifying the environment, they will achieve a better reproductive result. They do not do it on purpose. Rather, natural selection favoured individuals who actively intervened in their environment over those that did not, and, over the course of time, these ecosystem engineers emerged. Human beings, however, should know what they are doing and should be able to distinguish between useful and sustainable changes and those which, in the long term, will be harmful to the very species that set them in motion. But this awareness does not always emerge with sufficient clarity.

Sand, Bushes and Monsoons

Living organisms – plants, in particular – act on the climate in different ways. They modify the albedo, increase the transfer of water from the soil to the atmosphere and influence the carbon cycle. In 1975, US meteorologist Jule Charney, one of the founding fathers of modern meteorology, proposed a simple mechanism to describe how vegetation can influence the climate.

Take an area of desert in which there are a few bushes. Those bushes are generally darker than the sand beneath and so they increase the absorption of sunlight. If we suppose that, for some reason, part of that vegetation disappears, then the albedo (the fraction of sunlight that is reflected) will increase, resulting in the soil becoming colder. As a consequence, the temperature of the air in contact with the soil will decrease, making the stratification of the atmospheric column more stable. There will therefore be less vertical air mixing (weaker atmospheric convection) and, as a consequence, fewer storms, leading ultimately to less precipitation. As well as greater local stability, the change in atmospheric stratification will lead to a weakening of monsoons, with less movement of humid air towards the desert. Figure 6.1 shows a typical desert landscape in the Negev, not far from the Sde Boker university campus in Israel. We can clearly see both the darker colour of the vegetation in comparison to the sand, and the sparse distribution of the bushes, which, together with the biological crust, generate islands of fertility and humidity around them.

As plants in the desert are always struggling against the scarcity of water, a weakening of rainfall will lead to a further decrease in vegetation. In other words, the initial disturbance is amplified and could potentially lead to the collapse of the ecosystem. In the years following Charney's work, a series of studies carried out using climate models in which vegetation was artificially reduced (it is more complicated and less acceptable to carry out these tests in nature) confirmed this mechanism, demonstrating in particular how certain areas of the Sahel, in sub-Saharan Africa, can have two states: one that is entirely desert-like, arid and without precipitation, and one that is somewhat vegetated and more humid. We would certainly not find lush forests, but rather a distribution of sparse bushes like those we see today in the Negev.

Figure 6.1 Image of the Negev desert close to Sde-Boker in Israel, with its characteristic bushy vegetation that is darker than the sand and that also acts as an 'ecosystem engineer' alongside the biological crust and the isopods. Author's photo.

We see, therefore, that the climate of an entire portion of desert can be modified by the presence or absence of vegetation, and any disturbance to the plants can cause the temperature and precipitation conditions of the surrounding area to change.

The Lake beneath the Trees

As well as impacting the albedo, plants also profoundly modify the exchanges of water vapour between the soil and the atmosphere. The evaporation from liquid surfaces, such as the seas or the lakes, is responsible for the majority of vapour entering the atmosphere, forming clouds and, eventually, precipitation. However, the great expanses of vegetation also contribute significantly to the evaporation of water, often releasing into the atmosphere a significant percentage of the water that fell as rain and was absorbed by the soil. This phenomenon is known as evapotranspiration (the combination of evaporation and transpiration). In order to carry out their vital functions, such as photosynthesis, plants need to absorb water from the soil through their roots, using it and eventually transpiring it through the stomata in their leaves, which have to remain open to let in air (and therefore carbon dioxide) in order to perform photosynthesis.

On the surface of non-vegetated soil heated by the sun, the water can quickly evaporate from the top 5–10 cm, but the water lower down generally remains untouched. For several days or weeks after hard rainfall, the soil of non-vegetated desert areas remains damp at depths greater than 10 centimetres, precisely where evaporation is unable to take place effectively. In vegetated areas, however, the roots of the plants reach down to much greater depths, several metres even, and can therefore reach, use and transpire the water contained in these lower levels. Transpiration tends to be slower than pure evaporation from a sunny surface, but overall it transfers a higher quantity of water to the atmosphere.

Such a release of water from vegetation forms the basis for the local recycling of rainwater. In every sufficiently large portion of land, precipitation has two different origins: the condensation of vapour coming from afar (for example, from the sea) carried by atmospheric currents, and that of vapour that is 'evapotranspired' from the area's vegetation. In Amazonia, the contribution of water transpired by the forest becomes increasingly significant as you move downwind and into the forest. These observations have given rise to the saying 'rain follows the plough', which reminds us (albeit in an over-simplified and not always correct way) that an increase in vegetation can favour an increase in precipitation.

The role of local recycling also becomes crucial at times of summer drought in continental areas, as with the case of the unfortunate heatwave that occurred in France and Northern Italy in the summer of 2003. If, one summer, we have both a scant amount of water vapour from the ocean (the Atlantic in Europe's case) and a soil that is drier than normal because it has barely rained in the spring and snowed very little in the winter, then we start to lack both the water carried from afar and that recycled on a local level, and this can trigger a summer drought. In 2008, work by Italian ecologist Mara Baudena, climatologist Fabio D'Andrea from Paris's *École normale supérieure*, and myself, based on a conceptual description of the soil–vegetation–atmosphere system and motivated by the desire to understand what happened in 2003, revealed that, on a regional level, there were at least two possible scenarios: a 'normal' summer, relatively fresh and humid; and one that is drought-prone, very hot and dry. As we have said, this second state is favoured by the presence of an insufficiently moist soil at the beginning of the summer, and by scant vegetational covering.

Underneath the trees (or bushes or prairies), there is the equivalent of a lake contained in the soil's pores and in the surface aquifers, whose water is regularly transferred to the atmosphere through the actions of the vegetation. In this way, the released vapour enters the planetary water cycle – known as the hydrological cycle – before eventually returning to the soil and the oceans as precipitation. If we eliminate the vegetation and replace it with dirt roads or, worse, asphalt, we completely lose this essential component of the hydrological cycle and, in addition, we increase rain's surface run-off and erosion of the soil.

It is precisely the stabilization of the soil and its defence from erosion, together with the modification of the local climate, that led in 2007 to the idea of the 'Great green wall' in the Sahel, which was essentially the re-vegetation of a strip that would run through Africa for 8,000 kilometres from east to west, and which should help improve the environmental conditions of a region extremely at risk of drought and famine, modifying the climate, the hydrological cycle and erosion phenomena, and sand transport and desertification. A visionary idea, wanted and supported by the African populations with the backing of the United Nations, based on an alliance of different species of ecosystem engineers, including ourselves.

For all of these reasons, the role in the climate played by the thin layer that goes from the top of the vegetation to the bottom of surface aquifers has been increasingly studied over the last twenty years. Much research today is devoted to understanding the functioning of this 'Critical Zone', as it was defined at the beginning of this century by US researchers. This work was later supported by research programmes at the US National Science Foundation through the creation of a network of environmental observatories. The 'Critical Zone', composed of soil, rocks, water and living organisms ranging from bacteria and fungi to plants and animals, is where all the chemical, physical, geological and biological processes sustaining terrestrial ecosystems take place. It is the area where rock meets life; where rock weathering processes are intense; where exchanges of water, carbon and energy between the subsoil and the atmosphere occur. It is called the Critical Zone because it is fundamental to our survival, but it can be devastated by bad land management, deforestation, pollution or uncontrolled erosion. But it can be defended, supported and improved by work that focuses on the protection and intelligent management of the soil and the ecosystems. The book *Critical Zones*, published in 2020 and edited by French philosopher Bruno Latour and Austrian artist Peter Weibel, brings together scientific, artistic, philosophical and socio-political reflections on global changes and the Critical Zone.

The Bacterial Kingdom

Until now, we have spoken mostly about vegetation and a little about animals, such as beavers and isopods. But the great engines of our living planet have always been the bacteria that, alongside the Archea, dominated the scene and rendered the Earth habitable for 3.5 billion years before the impetuous development of the eukaryotes following the last Snowball episode. It is enough to remember the rise in the concentration of oxygen in the atmosphere, or the crucial role played by bacteria in the processes of rock weathering. Even today, the infinite multitudes of bacteria and Archea living on Earth are responsible for much of the activity that allows life on the planet, in the oceans and in the soil to thrive.

Unlike eukaryotes, including the multi-cell organisms that evolved very diverse physical forms and organization, bacteria have maintained their simple and not particularly variable cellular structure. However, they have developed an incredible variety of biochemical mechanisms that allow them to exploit almost any possible source of metabolic energy, managing to live in extreme conditions unthinkable for organisms with a more complex structure. Bacteria that live without oxygen, that use sulphur, that produce methane, are an apotheosis of the inventive chemistry of reactions that, for us humans used to aerobic life, often seem unusual. A great educational book by US microbiologist Paul Falkowski, *Life's Engines*, goes into detail about how bacteria have forged our world.

We can find clues as to how a bacterial world without oxygen – without photosynthesis, plants or animals – might have been by exploring the anoxic depths of the Black Sea. Here, we find both the fresh water from the great European rivers such as the Danube, as well as the water from the Mediterranean, saltier and therefore denser, that flows into it through the Bosporus. Upon encountering less dense water, that flowing from the Mediterranean sinks, creating two distinct layers: the water on the surface, fed by rivers and rains, which is less salty and lighter, and the denser, deeper layer from 100 to 130 metres down that is no longer in contact with the atmosphere. For this water, the stable stratification does not allow any mixing with the oxygen-rich water at the surface. As a result, the deepest parts of the Black Sea slowly lose their original oxygen and become akin to the oceans of the Archean. The bacteria that proliferate here are able to use sulphur for their metabolism, producing hydrogen sulphide (H_2S) and carbon dioxide as waste. A trip to the depths of the Black Sea can tell us how our planet's ancient oceans might have been.

Bacteria play a fundamental role in biogeochemical cycles, those cycles that cause the elements to circulate between the atmosphere, the ocean, the soil, living organisms and the Earth's crust. The nitrogen cycle, the iron cycle, the sulphur cycle, those of calcium and carbon are all dependent on the activity of the biosphere and, specifically, bacteria. We focus on those plants and animals that are closer to us and entirely visible to our eyes, but it is micro-organisms and fungi that drive the world forward. This is also the case for our food and drink – bread, cheese, wine, beer, to cite just a few – and the processes of decomposition and recycling of organic matter.

Gaia

We cannot end this chapter without referring to the most recent version of the system vision purported by von Humboldt, Wallace and Vernadsky on the living planet: the 'Gaia Hypothesis' put forward in the 1970s by English geochemist James Lovelock, and developed with the help of the great US biologist Lynn Margulis.

In essence, this hypothesis proposes that a planet such as the Earth is completely dominated by the actions of the biosphere, which has in some way taken control of the planetary dynamics and carries out a homeostatic action – meaning it is capable of reacting to external disturbances in order to keep conditions in the system stable and constant. The biosphere would therefore be able to keep the climate within the limits suited to the persistence of life itself. It is, in a certain sense, an extreme extension of the concept of niche construction, but in this case, it is the entirety of living organisms that build themselves an environment suited to their own permanence.

A classic example, developed by English scientist Andrew Watson, an expert in oceanography and the dynamics of the atmosphere, and by Lovelock himself, is a simplified model (a 'parable', according to the authors) called Daisyworld. Let's suppose that a planet exists in which there are two kinds of daisies: light ones that reflect the light of the star and cool the climate, and dark ones that absorb more light and warm the planet. Let's also suppose that the light daisies do better in the heat, precisely because they absorb less light, while the dark ones do better at lower temperatures. Now, if the star were to become brighter and therefore irradiate more, without daisies the planet's temperature would rise. But a rise in temperature would mean the light daisies would be favoured as a result and would expand their distribution. Consequently, the albedo would increase but the temperature wouldn't because the excess energy would be reflected. As a result, we have a mechanism capable of stabilizing the planet's temperature even if the star's luminosity were to change.

This parable was created to demonstrate how the biosphere might counter (within certain limits, at least) the changes caused by external forces such as solar luminosity. In reality, what is being hypothesized is not the role of the daisies, but the capacity of the biosphere to modify the composition of the atmosphere, the concentration of greenhouse gases and cloud coverage. This is undoubtedly true, but the point is whether this can effectively lead to a global and coherent homeostatic effect.

The Gaia Hypothesis has been widely debated, with some factions vociferously opposed to it, and discussions around it often moving beyond the boundaries of scientific reasoning. Many researchers are doubtful as to whether the biosphere would be able to react in such a coherent way to external changes, as the rules of evolution (random variation and natural selection) cannot have already acted upon a single superorganism that is not replicating, meaning there is an excessively finalistic flavour to the role of the biosphere as hypothesized by Lovelock. In order to exist, Gaia requires cooperative collective behaviour that is not easily reconcilable with the 'egotistical' behaviour that is often assumed to govern the dynamics of evolution by natural selection. But the question is fascinating and opens highly relevant research perspectives that cannot be cast aside. In recent decades, the work of British climatologist Tim Lenton and his colleagues explored in detail both the evolutionary aspects of Gaia and the possible role humans play in it (Lenton, 1998; Lenton and Latour, 2018), providing a good starting point for a quantitative understanding of the many facets of the Gaia Hypothesis.

7

Winds Up High and Currents in the Deep

Like the winds that powered the great sailing ships and the ocean currents that carried messages in glass bottles from far-away islands (and which, today, sadly carry more plastic bottles than anything else), the atmosphere and the Earth's oceans are in constant motion and form the circulatory system of our planet, helping it regulate itself. But what moves the air and the water, and what causes this incessant movement?

In order to understand, we must remember that the Earth receives a greater quantity of solar radiation per square metre in its tropical and equatorial zones than it does in polar areas, due to its spherical form and the fact that the axis of daily rotation is almost perpendicular to the plane of orbit.

This difference in energy received generates an imbalance, which causes the transport of heat from lower to higher latitudes in both hemispheres. It is precisely this imbalance that drives the oceanic and atmospheric circulation, the winds and the ocean currents.

Energy Received, Energy Re-radiated: A Question of Latitude

Since primary school, we have been taught that the Earth is round, and that at the tropics the Sun's rays arrive approximately at right angles to the surface, whilst towards the poles, the rays arrive at ever greater inclines. So, at the equator, the same quantity of incident solar energy is distributed over a much smaller area than at higher latitudes (north and south). As such, the energy that arrives in the upper atmosphere by unit of time and unit of surface area (let's say, per square metre) is much lower at higher latitudes than it is in tropical and equatorial zones.

Of course, not all of the energy that arrives in the upper atmosphere is absorbed by the planet's surface, as part of it is reflected by the clouds, and another by snow, ice and the light areas of the surface itself. To further complicate matters, this reflection (which we have referred to as albedo) varies with latitude but, for ease, we will ignore this particular detail for the moment.

If the axis of the planet's rotation were exactly perpendicular to the plane of its orbit around the star (parallel to the orbital axis), and the orbit were perfectly circular, there would be no seasons or differences in insolation between the two hemispheres. In this case, at the equator the Sun's rays at midday would always arrive perpendicular to the surface, whilst at the poles, the Sun would always be on the horizon bathing the Earth in an eternal half-light. Every latitude would always receive the same quantity of sunlight, which is divided symmetrically between the two hemispheres.

Instead, the Earth's rotation axis has a slight inclination with respect to the orbital axis, which today is around 23.5 degrees. In this way, even at the equator, the solar rays at midday are not always perpendicular to the surface. In the polar regions, beyond the Arctic or Antarctic Circles, the light conditions alternate between a perpetual day, with the Sun sitting above the horizon for many months (midnight sun), and a polar night during which the Sun cannot even be seen at midday. This inclination of the axis of rotation is responsible for the presence of the seasons. It is summer in a hemisphere when the inclination 'leans' towards the star and the sunlight arrives on the surface that is on less of an incline than the same latitude of the opposite hemisphere. What's more, the Earth's orbit is elliptic and so, during its journey around the Sun, the planet moves marginally away from or towards the star. During the summer in the Northern hemisphere, the Earth is slightly farther away from the Sun than it is during summer in the Southern hemisphere.

If we average it out over a year (the duration of a complete revolution by our planet around the Sun), we can see that the solar energy per unit of surface area received by the Earth has a maximum in the tropical and equatorial regions and a minimum in the polar areas. Receiving a different quantity of energy, each strip of latitude heats in a different way and therefore re-radiates a different total quantity of infrared energy. Here we must return to the albedo, as different latitudinal strips can have different reflective capacities – just think of the difference between the light sand of a desert and the dark rocks of a volcanic region. Therefore, it is not only incident radiation (determined by geometric and orbital factors) that we must consider, but the energy that is actually absorbed (and not reflected) by the surface. Figure 7.1 shows the energy absorbed by the Earth and that which is re-radiated, per second and square metre, according to latitude.

On a planet with no oceans or atmosphere, with limited possibilities for transferring the heat between different regions, every latitudinal strip would reach an equilibrium between energy absorbed and energy re-radiated. If, instead, we measure our planet's infrared emissions, we observe that, in polar regions, there is a greater emission than the energy absorbed, whilst at the tropics more energy is absorbed than re-radiated, as illustrated in figure 7.1. A situation of this kind might suggest a continual heating of low latitudes and a continual cooling of high latitudes, but this is not what actually happens. It is therefore necessary to understand which other mechanism is responsible for this local imbalance between energy received and energy emitted. And, here, we enter the world of the global circulation of the atmosphere and the oceans.

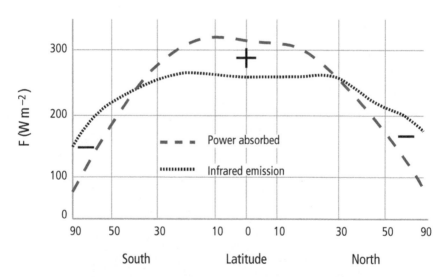

Figure 7.1 Power (energy per second in watts, W) by square metre absorbed by the Earth's surface (dashed line) and re-radiated by the Earth in the infrared (dotted line). Near the equator, the symbol '+' indicates that more power is received than re-radiated, while towards the poles, the symbols '−' indicate that more power is re-radiated than received from the Sun.

The Earth's Fluid Envelope

The density of the Earth's atmosphere rapidly decreases in an approximately exponential way with altitude. Similarly, atmospheric pressure, determined by the weight of the air column above, reduces with altitude, because there is increasingly less air above us, the higher we are above the surface. With altitude, the temperature, on the other hand, varies in a complex way, as shown in figure 7.2.

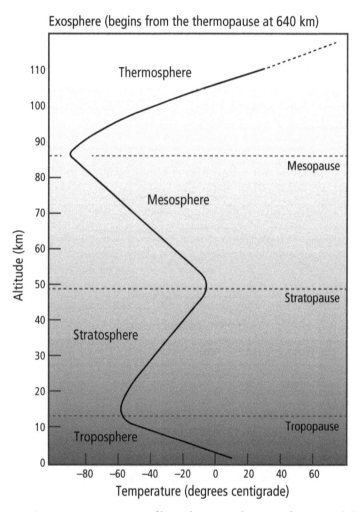

Figure 7.2 Average temperature profile in the atmosphere as a function of altitude.

The lower part of the atmosphere, known as the troposphere, is characterized by strong vertical convective motions that continuously mix the air together. Such convection is generated because the troposphere is heated from below. Solar radiation passes through the atmosphere and warms the planet's surface, which in turn releases heat into the atmosphere. The lower layer grows hotter and, according to the laws of gas, it also becomes less dense and tends to rise, causing the colder air that has stayed higher up to fall. Much like a saucepan of water that is heated from below, the fluid mixes and tends towards a neutral state, evenly redistributing the heat vertically.

However, unlike water in a pan, air is a gas and can therefore be compressed or expand more easily. As the air rises and encounters increasingly lower pressure, it expands and, again according to the laws of gas, cools. For such a compressible gas, then, the equilibrium distribution is not a state of constant temperature along the vertical, as with a saucepan of water, but rather a temperature that falls proportionately to altitude. For damp air, this leads to a vertical reduction of around 6 degrees per kilometre (though this can vary according to environmental conditions), while that of dry air is around 10 degrees per kilometre. If, due to being heated from below, the temperature decreases with altitude more rapidly than this, convective motions are triggered that mix the air vertically and tend to bring the situation back to a state of equilibrium. Hence the common occurrence on hot and sunny summer afternoons, when the soil has been sufficiently heated, of rather violent storms.

The troposphere contains most of the atmospheric mass and almost all of the water vapour, and the altitude it reaches varies with latitude: between 16 and 20 kilometres in equatorial zones and around 8 kilometres at the poles. At these altitudes, we find the tropopause, a layer of almost constant temperatures that precedes the stratosphere, which is situated higher up, reaching altitudes of around 50 kilometres. The stratosphere is more rarefied and generally contains very little water vapour. The temperatures in this layer rise with altitude, from around -60 °C in the tropopause to around -3 °C at the upper limit of the stratosphere. This rise in temperature along the vertical is due to the fact that, at altitudes of between 15 and 35 kilometres, we find the ozone (O_3) layer, which absorbs solar ultraviolet radiation and heats up, giving off heat in all directions. So, the stratosphere is heated from above and has a stable stratification in which the vertical convection motions are extremely limited or even absent.

Higher up again, we find other layers that are decreasingly dense: the mesosphere, which reaches an altitude of 85 kilometres and is heated from below by the ozone layer (here the temperatures drop as altitude increases), and the thermosphere, heated directly by solar radiation, where temperatures once again rise with altitude. Beyond this, we have the exosphere, with increasingly rarefied air and where the fastest molecules can easily escape into space. At the highest altitudes, the gases are for the most part ionized (meaning some electrons have been ripped from the atoms by the intense solar radiation) and their density becomes almost negligible at altitudes above 600 kilometres. To put this into perspective, the International Space Station sits at an altitude of 408 kilometres, while a normal passenger jet flies at altitudes reaching a maximum of 16 kilometres or just over, as was the case with the now-defunct Concorde.

An Atmosphere in Dynamic Equilibrium

Unlike the Moon or Mercury, the Earth has an atmosphere dense enough to support general circulation, which plays a crucial role in determining the planetary climate. Due to the intense solar energy absorbed in tropical zones, the air at lower latitudes is heated more and, as it becomes lighter, tends to rise upwards. The same thing happens at higher latitudes, but as the heating effect here is weaker, the air tends on average to rise less. Given that the density of the air diminishes as altitude increases, at a certain altitude (let's say 6,000 metres), the equatorial air will be denser than that at mid-latitudes and the poles. Such a difference in density produces a pressure imbalance that causes the air at higher altitudes to flow from the equator towards the poles, a little like water on a slope. But if the equatorial air rises and moves towards the poles, other air must move towards the equator at lower altitudes to compensate for the flow of air higher up. This ideal circulation would, therefore, form a large cell, known as a Hadley cell, which in each hemisphere has a side that ascends to the equator, a side that descends to the poles, a flow of air towards the equator at low altitudes and an opposite flow towards the poles higher up. Circulation from north to south – so, along the meridians – is called 'meridional circulation'. The panel at the top of figure 7.3 illustrates this hypothetical kind of circulation.

In the atmosphere of a planet that doesn't rotate, the dominant circulation would be formed by two large Hadley cells (one for each hemisphere) extended between the equator and the poles. The actual case, however, is more complicated. In the atmosphere of a rotating planet such as Earth, as well as the differences in density and pressure, there is another effect known as the Coriolis force. This is generated by the fact that we are rotating and, therefore, from a physical point of view, non-inertial. It is a little like what happens when we try to walk on a fairground ride that is rotating rapidly and we feel a force that moves us, making us deviate from the direction in which we are trying to move. It's the same thing for the Earth: with the atmospheric currents, the Coriolis force tends to deviate the airflow to the right in the Northern hemisphere and to the left in the Southern hemisphere, more so when the air movement is slow.

The combination of the differences in air density, the meridional circulation of heat between the equator and the poles, and the rotation of the planet makes hemispheric circulation in a single Hadley cell unstable. On the Earth, the movement of the atmosphere is, therefore, subdivided into 3 cells per hemisphere: the Hadley cell, which reaches latitudes of around 30 degrees, an intermediate cell (known as the Ferrel cell) situated between latitudes of 30–35 and 60 degrees, and a polar cell at latitudes of more than 60 degrees. At the equator and at around 60 degrees latitude (north or south), the air tends to rise (the ascending branch in the cells), while around 30 degrees and at the poles the air tends to fall (descending branch). It is no coincidence that at around 30 degrees there are the great deserts, because the air that falls is warmed and becomes drier, reducing precipitation. But we will talk about this more in chapter 9.

The Hadley cell and the polar cell are characterized by air that moves towards the equator close to the ground and a return flow at altitude towards the pole. In contrast, in the Ferrel cell, the air on the ground moves from the equator towards the poles, and vice versa for the return flow at altitude. The Coriolis force shifts these air flows in a lateral direction. In the tropical zones of the Northern hemisphere (the Southern hemisphere is symmetrical), the Coriolis effect causes

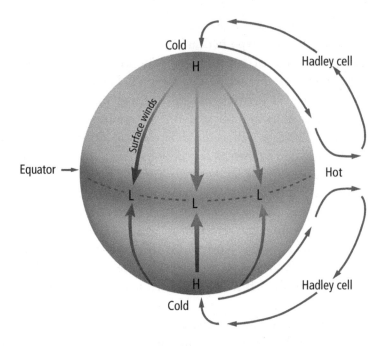

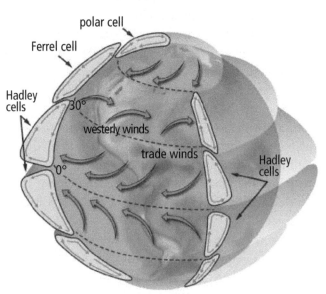

Figure 7.3 General circulation of the atmosphere: above, in the case of a single Hadley cell (in the absence of planetary rotation); and in the actual case of the Earth, below.

the winds to rotate in a south-west direction. These are known as trade winds. At mid-latitudes, instead, the low-level flow towards the pole is shifted towards the north-east (the westerlies, winds that arrive from the west). This is the reason why, in Europe, atmospheric perturbations tend to arrive from the Atlantic, carried by the westerlies, while in America they arrive from the Pacific.

The Hadley cell and the polar cell are relatively stable, and easily recognizable both in satellite images and in the data provided by measuring networks on the ground. The Ferrel cell, on the other hand, is highly unstable and can only be identified if we take an average over months or years. The circulation at mid-latitudes is extremely variable from day to day, shifting from good weather conditions to intense precipitation, with temperature changes of several degrees. This is the dynamics of 'extratropical cyclones' where hot and cold fronts and strong winds alternating in all directions agitate the weather at mid-latitudes, making it particularly lively.

In different climate conditions, the extension of the three hemispheric cells can change – significantly, even. The Hadley cell, in particular, can expand or contract, influencing the meteorological regime and precipitation of entire regions. In the climate changes currently under way, there is evidence of modifications to the extension of the Hadley cell, and even reconstructions of past climates reveal a shifting of arid or rainy zones caused by changes in atmospheric circulation.

Jet Streams and Storm Tracks

In the upper troposphere, just beneath the tropopause at altitudes between 10 and 16 kilometres, we find some very strong and localized currents, with wind speeds that can reach and sometimes exceed 180 km/h. Each hemisphere has two of these currents, with the winds always blowing from west to east. They are called jet streams, and can be found roughly on the boundary between the Hadley and Ferrel cells (the subtropical jet), and between the Ferrel and polar cells (the polar jet). These currents oscillate, their shape shifts and they are traversed by waves (named Rossby waves after the Swedish meteorologist who first identified them) that generate meanders and contortions, causing the direction of the wind to vary in time and space.

The jet streams were discovered in the first half of the last century by meteorologists and pilots. It was Japanese meteorologist Wasaburo Oishi and US pilot Wiley Post in particular who first noticed the existence of very strong winds at altitude. After this, warplane pilots flying very high were able to experience directly the strength of these streams, which accelerated or slowed their movement depending on the direction of flight.

The polar jet plays a particularly important role in determining the trajectory of meteorological perturbations at mid-latitudes associated with the dynamics of Rossby waves. By tracing the movement of storms, both tropical and extratropical, we can define the 'storm track', which is a more or less localized area through which the perturbations will move. The extratropical branch of this path is strongly influenced by jet streams, and the entire system is sensitive to global changes in climate, as we will see again for the Arctic in chapter 16.

It is important to emphasize how the structure of three cells and two jet streams per hemisphere is not universal. For planets with a slower or faster rotation period than the Earth, with a different radius or a different tilt on the rotational axis with respect to the orbital plane, the structure of the atmospheric circulation can be very different. Venus, for example, turns much more slowly than the Earth and has Hadley cells that are much larger than those found on our planet. Jupiter, an intriguing laboratory for anyone studying planetary fluid motion, has a circulation characterized by extremely strong currents alternating with vortices of all sizes. Here we are entering the wonderful world of other planets in the solar system and exoplanets, those thousands of planets that have been discovered around other stars over the last twenty years. These distant bodies provide (or will provide) a unique opportunity to study the climate and dynamics of atmospheres and oceans in conditions very different from those currently found on Earth, helping us develop a 'comparative climatology' and gain a deeper understanding of how the climate functions. We just have to think of that excellent film *Interstellar* to have a qualitative idea of how the climate might be on other rocky planets similar to Earth.

Planet Ocean

Around 70 per cent of the Earth's surface is occupied by seas and oceans, leading some to believe our planet should be called not Earth, but

Ocean. The oceans have an average depth of around 4,000 metres, but in some points reach much greater depths, including up to 11,000 metres in the Pacific Ocean's Mariana Trench.

Contrary to what we see with the atmosphere, the oceans are warmed from above because solar radiation is absorbed almost entirely by the 100 or 200 metres of water closest to the surface, known as the euphotic zone. This zone is rich in light and it is where phytoplankton and aquatic plants carry out photosynthesis, as well as being the location of most of the processes in the marine ecosystem. Due to warming, the surface water is lighter, and so floats on deeper, denser water. The ocean, therefore, is almost entirely characterized by a stable stratification with vertical motions generally located in specific convection regions or along temperature fronts, and a more superficial layer that is around 100 metres deep and continuously mixed by the wind.

What's more, seawater is salty due to the many dissolved salts that have ended up in the sea through the erosion and chemical weathering of the rocks on the continents. At high latitudes, superficial salinity (around 33‰) is lower than that in tropical regions, where it can be higher than 36‰ due to intense evaporation and scarce precipitation, before then dropping in equatorial areas due to high levels of precipitation. At depths greater than a kilometre, the salinity tends to be roughly constant around just under 35‰ (parts per thousand). In the 1,000 metres nearest the surface, however, the salinity profile can vary according to latitude. Figure 7.4 illustrates the vertical stratification of temperature, salinity and density in average open ocean conditions. The region between the superficial mixed layer and a depth of 500–700 metres is called the thermocline, the zone where the temperature varies most rapidly with depth.

The ocean, however, is not a calm pool in which the deep waters stay at the bottom of the sea without exchanges with the atmosphere. If this were the case, the deep ocean would quickly become anoxic (deprived of oxygen) as has happened at times in the Earth's past. In some, very localized points, however, around both the Antarctic and the Arctic, the surface water cools so vigorously due to the very cold air and strong winds that it can become heavier than the water lower down and quickly sinks, in some cases to the ocean floor. In these deep regions, the water then moves towards tropical and equatorial latitudes where it slowly

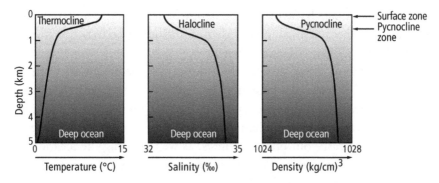

Figure 7.4 Vertical profile of temperature, salinity and density in typical ocean conditions. *Source:* © 2004 Pearson Prentice Hall, Inc.

re-emerges and mixes with the surface water. This is called the thermohaline circulation, the great conveyor belt of oceanic heat, which we will look at in more depth shortly.

Interestingly, it is worth remembering that one of the very few points where water 'sinks' outside the polar regions is found in the Mediterranean, in the Gulf of Lion offshore from Marseilles. Here, during the winter, the strong north-westerly wind (the mistral) encourages evaporation and cools the sea's surface, making the water saltier, colder and, as a result, denser. The surface water then sinks into the sea, generating a deep circulation in the Mediterranean that is the same, albeit on a smaller scale, as that found in our planet's oceans.

Oceanic Currents

The water in the ocean is extremely mobile and participates in the transport of heat from the tropics to the poles, working alongside the atmosphere, with which it actually forms a closely connected system, as described in the book *Ocean Circulation Theory* by oceanographer Joseph Pedlosky of the Woods Hole Oceanographic Institution in the US.

One kind of circulation is thermohaline (thermal and saline), which we mentioned earlier. The surface waters, cooled in the polar regions, sink and move towards the equator and partially resurface there, mixing once more with the water from lower latitudes before then returning to the poles on the ocean's superficial layer. In the Atlantic, the deep waters from the Arctic partially cross the equator and move towards the Antarctic. The American oceanographer and climatologist Wallace Broecker was a pioneer in the study of the thermohaline circulation and coined the term 'global conveyor belt' to indicate this movement by the water, which slowly unfurls throughout all oceans without any interruptions. Typically, a water particle takes a few hundred years to 'take a full turn' of the entire conveyor belt, with a typical average speed of around 1 centimetre per second. Figure 7.5 shows a simplified diagram of this circulation, as viewed from the South Pole, and highlights the continuity between the different oceans. In truth, the actual circulation is much more irregular and variable than this simple diagram, as we would expect of a turbulent flow.

A second kind of oceanic circulation is linked to the direct action of the wind, which generates currents across all oceans. As we have seen, the wind on a planetary scale comes from the differences in warming between the tropics and the poles. It blows on the ocean water and

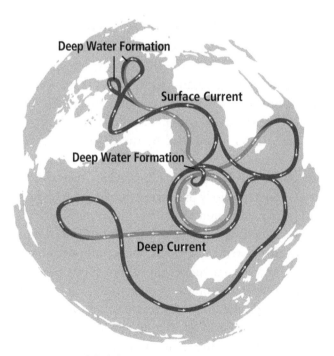

Figure 7.5 A simplified diagram of oceanic thermohaline circulation.

moves it, generating circular movements in all ocean basins. This kind of circulation is referred to as wind-driven, and reaches depths ranging from about 1,000 metres at low latitudes to much greater depths as it moves towards the poles, where the stratification is less pronounced.

In every ocean basin, two enormous gyres (closed circulation systems) are formed: one that is subtropical, located between the equator and the latitude where the westerly winds are more intense (around 50° north or south), and one that is subpolar, located between these latitudes and the poles. In the Northern hemisphere, the subtropical gyre turns clockwise and the subpolar gyre turns anti-clockwise, while the opposite occurs in the Southern hemisphere. In each of these gyres, due to the rotation and spherical shape of the Earth, the currents intensify at the western margins of the basins, as is the case with the Gulf Stream in the North Atlantic and the Kuroshio current in the North Pacific. In 1948, US oceanographer Henry Stommel, one of the founding fathers of the study of oceanic flows, developed a physical and mathematical explanation of this characteristic, laying the foundations for the quantitative under-standing of general ocean circulation. The circulation forced by the wind is decidedly faster than the thermohaline, with one water particle taking on average just a few years to complete an entire turn of a basin.

Figure 7.6 schematically illustrates the wind-driven circulation forced by the wind in the different ocean basins. The currents were discovered when the speed of water was measured by oceanographic vessels using chains of current metres, or by commercial ships travelling the world by sea. Further information is provided by the trajectories of free-moving ocean floats on the surface or at depth, which transmit their position and the main characteristics of the water in which they find themselves. Satellites are also used, precisely measuring the elevation of the sea surface and comparing it to a situation of static equilibrium, which allows us to ascertain the water's average speed using the formulas of fluid dynamics.

It is interesting to remember that some information on oceanic currents has also come from rubber ducks and sports shoes. In 1990, a cargo ship in the Pacific lost some 60,000 gym shoes in a storm. In 1992, another ship in the Pacific lost more than 28,000 rubber ducks. In the following years, shoes and ducks were found in the most disparate of coastal locations, providing information on marine currents. The book *Moby Duck* by journalist Donovan Hohn tells this story, which is not as

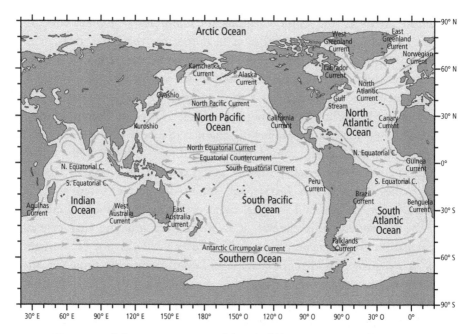

Figure 7.6 Schematic structure of the wind-driven ocean circulation.

fun as it might initially seem. The quantity of plastic released into the oceans is indeed terrifying, and has generated floating plastic islands of colossal proportions in each of the main oceans. This is, in all senses, one of the largest anthropic impacts on our planet's environment.

The thermohaline circulation and that caused by the wind are obviously not independent of one another. In many cases, the intense currents caused by the wind on the western side of the ocean basins, such as the Gulf Stream, are also the branch of surface return flow in the thermohaline circulation. In this way, the oceanic motions become a single system characterized by many spatial and temporal scales that are active at the same time and strongly connected to atmospheric circulation.

A Question of Transport

General circulation in the atmosphere and the ocean is 'set in motion' by the differences in insolation, and therefore temperature, pressure and density, between the tropical regions and the poles. The main effect of this circulation is the transport of heat towards the polar areas, with the consequent reduction of those same differences in temperature that generate the fluid movements. Thus, we arrive at a situation in which differences in temperature and the intensity of the currents reach a dynamic equilibrium, regulating one another. In very different climate conditions – for example, following changes in albedo or the greenhouse effect – this equilibrium can be modified, with the differences in temperature between the various latitudes being potentially bigger or smaller than they currently are (as we will see in the next chapter) and the intensity of the heat and material transport being profoundly altered.

Before we end this chapter, it is worth remembering that general circulation in the atmosphere and the ocean should not be thought of as a regular flow that is constant and continuous over time. On the contrary, variability over time and space dominates the fluid motions, generating turbulence on all scales. At temperate latitudes, the Ferrel cell only emerges as the average of an otherwise wildly variable weather – day after day, cyclones follow anticyclones, rain follows clear skies in a continual cycle. Towards the poles, the Arctic and Antarctic atmospheric subpolar vortices are constantly disturbed by huge planetary waves that bring ice-cold air to the south and warm air towards the poles. In tropical regions, the relative regularity of the Hadley circulation is often interrupted by powerful hurricanes (or typhoons or tropical cyclones as they are known in other parts of the world) that generate impetuous winds and churn up the marine waters. And we shouldn't think of those intense ocean currents on the western margin of the basins, such as the Gulf Stream, as slow and regular rivers, but as complex structures that are variable, dynamic and always disturbed by meanders and great vortices that detach and move around the ocean, sometimes for years. This is the intense 'mesoscale' oceanic turbulence, capable of mixing the ocean water and controlling the marine ecosystem, and studied (for example) by James McWilliams at the University of California in Los Angeles and Annalisa Bracco at the Georgia Institute of Technology in Atlanta, along with numerous other colleagues. In short, ours is a dynamic planet with a lively and changeable circulation that acts as a heat distribution system and regulator of the Earth's climate.

8

The Big Heat

After the wide-scale oscillations that followed the Earth's emergence from the Snowball, starting from the Jurassic period around 200 million years ago, surface temperatures on Earth slowly began to rise. A maximum peak was reached just over 50 million years ago in the period known as the Eocene, when the Earth was, for the most part, ice-free. However, around 55 million years ago, this relatively steady growth was interrupted by a particular moment in the history of the climate characterized by a violent rise in temperature that was rapid and of limited duration, in a geological sense at least. This sudden event that shook the environment and the ecosystems is known as the PETM: Paleocene–Eocene Thermal Maximum. We will start by looking at conditions during the Eocene, before concentrating on the PETM and its causes.

An Equable Climate

During the Eocene, the average temperature on the Earth's surface was around 10 °C higher than it is currently (estimates vary between 7 and 14 degrees higher depending on the reconstruction method used). Starting around 60 million years ago, the information available on the climate becomes more abundant and precise, and the analysis of oceanic sediments is able to provide us with an accurate timeline and fairly detailed indications on the climate conditions. Figure 8.1 shows the most famous temperature estimates of the last 60 million years, named the Zachos curve after US oceanographer and paleoclimatologist James Zachos, the coordinator of the work that made this reconstruction possible.

This figure clearly shows a period of elevated temperatures during the Eocene and the temperature peak corresponding with the PETM. After the Eocene, temperatures began to fall until reaching our current ice age, which began around 3 million years ago. But we will discuss this later on.

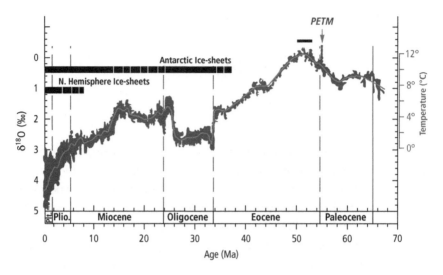

Figure 8.1 The Zachos curve. Time runs from right to left. On the left-hand vertical axis, we see the variation in the isotopic ratio between oxygen-18 and oxygen-16, while on the right-hand vertical axis we find the estimated temperature differences compared with those found today. The two horizontal bars on the left show the presence of polar ice caps in the Antarctic (above) and the Arctic (below), with the solid line indicating permanent ice caps and the dotted line indicating periods of temporary or fragmented presence. At the bottom are the geological periods. Taken from James Zachos et al. (2001).

The Eocene had a very hot climate but, even more importantly, there was only a minor difference in temperature between the equator and the poles. It was an equable climate, more homogeneous and balanced. The paleoclimatic data indicates that temperatures in the tropical and equatorial zones were 4 or 5 degrees higher than they are today. On the other hand, at high latitudes the warming was higher than 10 or 15 degrees more, and in some continental regions it could have reached up to 20 degrees more than current temperatures. It was an extremely hot world by our standards. One in which the water cycle was, with every probability, much more violent, and which saw an incessant series of hurricanes that were not confined to the tropics.

What's more, fossils tell us the same story as the information provided by sediments and isotopic analysis. In the North Slope of Alaska, on the Spitzbergen island in Svalbard and on Ellesmere Island (Canada), at around 78° north both then and today, where we now find only Arctic tundra, there were once forests of conifers and ginkgo. On Ellesmere, researchers have found fossils of monitor lizards, tortoises, alligators and animals similar to the crocodiles that today live only in regions with average annual temperatures above 15 degrees and with minimums no lower than 5 degrees. There was no (or very little) ice, and ecosystems that we find today in temperate or tropical climates thrived in the Arctic. Moving towards the South Pole, we find a similar situation, with temperatures already much higher before the Eocene. Paleontologists have found fossils of dinosaurs from the Jurassic and Cretaceous periods in Antarctica, as well as in Siberia and the Yukon in the Northern hemisphere. These fossils include species that were most likely cold-blooded and therefore required a relatively warm environment in order to survive. This is precisely the point: it is not enough to say that dinosaurs were close to the poles in order to conclude that the climate was hotter, but it is essential to find (as has happened) fossils of species that were presumably cold-blooded and therefore incapable of thriving in a frozen environment.

Winds and Hurricanes in the Equable Climate

But which climate processes were responsible for the high temperatures during the Eocene? Isotropic analysis of the sediments indicates that the concentration of carbon dioxide in the atmosphere was probably much higher than it is today, almost certainly more than 1,000 ppm, compared with around 417 in 2020, even if estimates do vary quite significantly. The problem, however, is that even such elevated levels of carbon dioxide cannot alone explain the greater rise in temperature at the poles compared to that at lower latitudes. This is why we need to identify other mechanisms – a task with which paleoclimatologists have been grappling for the last thirty years, without entirely satisfying results to date. However, by attempting to solve this mystery, we can learn a lot about the climate and the mechanisms of amplification and control that are still active today.

One possibility involves the intensity of heat transport from the tropics to the poles, looking back at what we saw in chapter 7. If the transport had been much more efficient than it is today, then the excess heat accumulated at the equator could have been rapidly distributed to higher latitudes, warming them vigorously.

There is, however, the problem of how this heat transport could have actually happened in practice. In the simplest of cases, the heat transport from a hotter zone to a colder one depends on the temperature difference between the two regions: the smaller this difference, the less intense the transport. Therefore, if some perturbation of the general circulation had at a certain point triggered a powerful transport of heat, reducing the differences in temperature between the equator and the poles, then this same reduction would have subsequently weakened the transport, favouring a return to the previous conditions. Unless, that is, there were other, more complex processes that could have sustained a heat transport that was more intense than usual for many millions of years.

One such mechanism, put forward by US meteorologist Brian Farrell in 1990, suggests that during the Eocene, the Hadley cell (which, as we have seen, governs atmospheric dynamics in the tropics) extended towards higher latitudes, dominated by winds moving from west to east. However, the drivers that could have caused this to happen are not clear and this interpretation has now substantially been abandoned.

Another mechanism was suggested by US meteorologists Robert Korty, Kerry Emanuel and Jeffery Scott in 2008, who hypothesized a greater intensity and larger number of tropical storms and hurricanes in the warm conditions of the Eocene. These storms mix the ocean water, bringing deeper, colder water to the surface, which is then heated rapidly by the Sun. After this, the oceanic circulation moves warm water towards polar latitudes, which are in turn heated. But more recent work has shown that the situation is more complicated, and that it is not necessarily true that hurricanes can increase heat transfer towards the poles. For now, at least, the explanations based on a rise in heat transport do not provide the results we had hoped for.

Is the Answer in the Clouds?

There are other possibilities for explaining the warming at higher latitudes during the Eocene that do not require any modification of the global circulation. One line of investigation considers the role of the clouds, both those high in the stratosphere and the convection clouds (storm clouds) in the troposphere.

In the stratosphere, there is very little water vapour due to the low temperatures, as this comes mainly from the upward displacement of air hailing from the lower atmosphere. Even today, stratospheric polar clouds can form, but they are very thin and make only a small contribution to the greenhouse effect and localized warming in the Arctic and Antarctic regions.

In 1998, US climatologists Lisa Cirbus Sloan and David Pollard put forward a theory that, due to the high temperatures of the Eocene and the preceding periods, many more swamps and damp zones developed, resulting in increased methane emissions. Once it reaches the stratosphere, methane can oxidize, producing carbon dioxide and water. And so, here we have a possible mechanism for generating water in the upper atmosphere and increasing the impact on the greenhouse effect from polar stratospheric clouds. It is still a greenhouse effect, but this time it is mediated by methane (as it was for hundreds of millions of years) and not directly due to carbon dioxide.

The idea of a greater concentration of methane has, however, been highly criticized, particularly because methane has a half-life of around 7 years in today's atmosphere but the extreme warming of the Eocene lasted many millions of years. It would therefore require an enormous and continuous emission of methane from the swamps, probably far greater than is actually possible or actually happened. Various researchers later explored other paths that might lead to an increase in the presence of polar stratospheric clouds that are not necessarily linked to the presence of methane, but rather to the warming of the stratosphere and changes in atmospheric circulation at high altitudes. The idea of polar clouds, therefore, though not fully developed, remains an interesting possibility.

Clouds also form the basis for the explanation put forward in 2008 by climatologists Dorian Abbot and Eli Tziperman – though this time in the troposphere. According to these researchers, warming in the Eocene was almost exclusively localized in the polar regions, caused by a rise in the concentration of carbon dioxide in the atmosphere amplified by feedback processes on a regional scale. By generating a rise in temperature, the greenhouse effect led to the melting of the polar marine ice in a way not too dissimilar from what is happening in current times.

In this hypothesis, without ice to act as a lid, evaporation from the sea would have caused a rise in the concentration of water vapour in the polar atmosphere, which as it heated would contain more and more vapour, generating convection (storm) clouds as happens at lower latitudes. This process could have caused an amplification of the warming due to the strong greenhouse effect of vapour and the clouds. In addition, the cooling effect potentially associated with the reflection of solar radiation from the clouds was absent, because these processes happened mainly during the polar night when only low or no solar radiation was received. At the moment, we do not know whether this explanation is the right one (at least in part) but it is certainly very promising. And it sets alarm bells ringing about the possible consequences of the disappearance of Arctic sea ice witnessed in recent decades.

Finally, it is important to remember another element of this complicated picture: the role of vegetation in polar areas. In the Eocene, polar ice was almost entirely absent, meaning there was soil and vegetation in both Greenland and on the Antarctic continent. The vegetation would without doubt have been darker than the ice and would therefore have absorbed more solar radiation in the summer, warming itself and the environment, as we saw in chapter 6: a Charney mechanism for the polar regions instead of the deserts. This explanation was put forward by US paleobiologist Garland Upchurch and colleagues in 1998, and even if it is not the main mechanism for polar warming in the Eocene and the Cretaceous, it is nevertheless a relevant amplification factor. Once again, we have an example of how the biosphere can influence the climate on a continental, if not global, scale.

So Just How Much Carbon Dioxide Was There in the Eocene?

It is by no means simple to determine the concentration of carbon dioxide in the atmosphere during a period so distant from our own. As we saw earlier, the majority of estimates are based on isotopic methods, particularly on the relative abundance of carbon-13 and carbon-12, and, more recently, the stable isotopes of boron, boron-10 and boron-11, measured in the calcareous shells of marine organisms. Other information comes from the relative presence of magnesium and calcium ions in seawater, which can be reconstructed through an analysis of evaporites, rocks that are formed when there is a massive evaporation of seawater in enclosed, relatively shallow basins. The analysis of the dimensions and the structure of stomata in fossilized leaves also provides an indirect estimate of the concentration of carbon dioxide in the atmosphere: the greater the concentration of CO_2, the greater and more profound the stomata.

When considered as a whole, all of this indirect information tells us that the concentration of carbon dioxide in the atmosphere during the Eocene was much larger than it is today, probably by more than 1,000 ppm. But most climate models used to explain polar warming in that period require even greater levels of CO_2 concentration, up to 4,000 ppm or more. Evidently, there is still something missing from climate models, something that would be capable of reproducing the Eocene climate without requiring enormous quantities of greenhouse gases in the atmosphere.

Amplification mechanisms, such as the effect of stratospheric or convection clouds described earlier, can massively increase warming, even with relatively low levels of greenhouse gases. In this case, carbon dioxide would fire the starting signal but other climate processes would take over, reinforcing that initial stimulus. If this were the case, however, our attention to the current rise in the concentration of greenhouse gases should be higher, because, even at not particularly elevated concentrations, these gases could trigger feedback processes pushing full speed towards a climate like that of the Eocene. In any case, not least because of the potential importance of these conditions to our own future, our understanding of the 'equable' climate of the Eocene will occupy researchers for years to come, together with the great and sudden warming event known as the Paleocene–Eocene Thermal Maximum, which we will now explore.

A Big, Sudden Heat

The most dramatic climate event after the Snowball was the thermal maximum reached between the Paleocene and the Eocene, just over 55 million years ago – the PETM we referred to at the beginning of the chapter.

This period of extreme heat, first identified by US marine geologists James P. Kennett and Lowell D. Stott in 1991, lasted around 200,000 years. The data indicates that it took 10,000 years for the temperatures to rise by between 5 and 8 degrees on the already elevated levels of the late Cretaceous, again with a much greater rise at the poles than in the tropical zones. During this 'rapid' event (in a geological sense), the temperatures rose by around 0.8 degrees every 1,000 years. To put this into perspective, just think that, over the last century, global temperatures rose almost 1 whole degree in 100 years, therefore with a speed almost 10 times higher and extremely unusual for the natural variability of the planet's climate. But we will go into this in more depth later on.

Going back to the PETM, the rise in temperature was accompanied by significant changes in oceanic circulation and the water cycle. The ecosystems also responded to the warming, and the sediments deposited at that time show an abundant contribution of carbon of organic origin, as shown by the sudden change in the isotopic ratio between carbon-13 and carbon-12. The entire weathering cycle of surface rocks, dependent on both the temperature and the acidity of precipitation, was profoundly modified. What's more, the ocean underwent a process of acidification caused by the rise in carbon dioxide in the atmosphere.

What actually caused the PETM event is still the subject of scientific discussion. There were almost certainly several contributing factors unleashed by an initial and significant release of greenhouse gases that led to a rapid increase in the atmospheric concentration of carbon dioxide. It is estimated that the warming measured during the PETM is compatible with the emission of a quantity of carbon between 2,000 and 6,000 gigatonnes (Gt), which happened over many thousands of years. We should also remember that the anthropic emissions of the last few years are the equivalent of more than 10 Gt of carbon per year, so, at the current rate, it would take less than 200 years to reach the critical threshold that triggered the dynamic observed in the PETM.

The beginning of the entire warming process could have been rooted in a series of powerful and prolonged volcanic eruptions associated with the movement of the plates and the expansion of the Atlantic Ocean basin. But, as we have said, these volcanic emissions alone do not seem to be enough to explain the enormous rise in temperature.

In this case also, it is necessary to identify other mechanisms capable of amplifying the initial stimulus. Researchers have analysed various possibilities, verifying their compatibility with the data provided by the sediments. Here, we will cite just a few of the most promising hypotheses. The first, proposed by marine geologist and paleoceanographer Gerald Dickens and his team in 1995, is based on the release of greenhouse gases by the methane hydrates contained in marine sediment. Gas hydrates are ice-like crystalline solids formed by a network of water molecules that 'trap' the molecules of a light gas and are stable at low temperatures and high pressures. If the temperature rises, the gas contained can be rapidly released and return to the atmosphere. In the case of methane hydrates (also called 'burning ice' because it looks like ice but if you put it near a flame it ignites), the methane released can contribute to a powerful rise in the greenhouse effect. The idea is, therefore, that an initial rise in temperature in the atmosphere and the ocean water, caused by the effect of carbon dioxide released by the volcanoes, was massively increased by the destruction of methane hydrates and the consequent release of this gas.

Another very promising idea was proposed in 2021 by a group of researchers coordinated by US climatologist Robert DeConto. According to this hypothesis, even if the polar regions were virtually free of ice in the period preceding the PETM, they were nevertheless partially covered in permafrost, the perennially frozen soil that we find today in polar regions and high in the mountains. The permafrost contains large quantities of organic matter that does not decompose because it is frozen. However, if the temperature rises and the permafrost thaws, the effect is much the same as when you leave your freezer unplugged: the organic content quickly decomposes and, in doing so, releases great quantities of methane into the atmosphere, thus amplifying the initial warming even more.

As always, extreme climate events have many contributing factors and, most importantly, are characterized by a chain of processes that can increase the effect of the initial perturbation. There is still much to discover about the PETM, which, together with the Snowball, is perhaps one of the most intriguing moments in the dynamics of our climate. Furthermore, the PETM event is of immediate interest because it can help in our understanding of the climate that awaits us, as the quantities of carbon initially released by volcanoes during the Eocene are comparable to anthropic carbon dioxide emissions. On top of this, the same amplification mechanisms identified for that distant event (the release of methane from methane hydrates and the thawing of the permafrost) are still active and powerful today.

9

Rain, Snow and Clouds: The Planetary Water Cycle

The rain fell incessantly every day. The ground that had once been parched was now swollen with water, and little streams ran down the things and people in the street. The wet roofs shone and poured impetuous cascades into drainpipes, while clouds of vapour rose from the asphalt and wet clothes. Who knows where all that water had come from, which celestial paths it had followed, bringing messages from far-away lands and even more remote seas.

The Triple Point

The Earth's atmosphere contains small quantities of water in all three of its states: vapour, liquid and solid (ice and snow). This is one of the fundamental characteristics of our planet: that the conditions of temperature and pressure at the surface and in the troposphere are never too distant from the 'triple point' of water, the point at which its three states can coexist. This means that minor variations in temperature or pressure can cause water to go from being a vapour to a liquid or from a liquid to a solid or vice versa, or it can jump the middle step and go straight from being solid to a vapour, or even solidify itself directly from vapour.

It is worth remembering the distinction between vapour and gas. They are the same state, but vapour is found below what is known as the 'critical temperature'. In these conditions, the coexistence with a liquid state is possible and a vapour can be condensed both by a drop in temperature and through the variation of pressure in an environment with a constant temperature. Above the critical temperature, however, the only possible state is a gaseous one, while at extremely low temperatures only the solid phase (ice) can exist. Liquid water, crucial for biological processes, can be found in the intermediary zone between the two extremes of gas and perennial ice. Not by chance, one of the conditions that is consistently required for any possible existence of life on other planets is the presence of water in a liquid state.

With every change in state, a certain amount of heat is released or absorbed. This is known as 'latent heat'. Water molecules generally have more kinetic energy (they move a lot) in a vapour or gaseous state. This is the most chaotic state as there are no longer close bonds between the molecules, which can, therefore, move around freely, continually bumping into one another. In the liquid phase, the molecules can still move but they exercise forces of attraction or repulsion on one another, and are therefore bonded. Kinetic energy here is lower and so the liquid phase is, as it were, better behaved than vapour. The solid state is the most orderly, the molecules are frozen (literally in this case) in a crystalline network and can only vibrate slightly around their equilibrium position.

In the change from the disordered to the more orderly, as happens in the condensation of vapour into a liquid or the solidification of liquid into ice, heat is released because the molecules slow down their movements and their kinetic energy therefore decreases. In contrast, in evaporation or melting there is an absorption of heat because the molecules require additional energy to break the chains that bind them. In this way, the entirety of Earth's atmosphere becomes a thermal machine in which impressive quantities of heat are exchanged and transported through changes in state. If anyone were interested, it is possible to find a splendid introduction to thermodynamics in a small 1937 book of the same name (*Thermodynamics*) by Enrico Fermi, the great Italian physicist who founded the International School of Physics in Rome and was awarded the Nobel Prize in 1938, shortly before he was forced to emigrate to the United States by the Fascists' wretched racial laws.

Vapour and Droplets

The transfer of heat through evaporation and condensation is one of the basic processes by which the Earth's surface, heated by the Sun, releases heat into the atmosphere. As well as the emission of infrared radiation and the cooling of the soil, which gives off heat when it comes into contact with the colder air around it, the evaporation of water from lakes and seas and transpiration from plants removes heat from liquid surfaces and the soil before returning it to the atmosphere. The vapour formed close to the surface moves higher up, together with the warmed air that, having become lighter, also rises, generating turbulent convective motions. On average, more than 80 W/m² are given off by evapotranspiration, around 20 W/m² by the heating of the air in direct contact with the surface, and around 400 W/m² by infrared radiation, of which 340 W/m² then return to the surface because of the greenhouse effect.

The air that rises encounters ever-diminishing pressures and, as per the gas laws, it expands and cools. Here another fundamental law of thermodynamics comes into play, described by the Clausius–Clapeyron equation we looked at in chapter 3. The maximum quantity of water vapour that can be contained in a defined volume of air generally depends on the air temperature. The hotter the air, the greater the quantity of vapour that can be held. Relative humidity, used in meteorology, is defined by the ratio between the amount of vapour present in the air and the maximum quantity that can be contained.

When the temperature of a volume of air decreases – due, for example, to a rise in altitude and consequent expansion of the air, or because it mixes with very cold air or comes into contact with a cold surface – the maximum quantity of vapour that can be contained also rapidly declines. At a certain point, when the vapour that was initially present is in excess of what can be contained, the air becomes saturated and part of this vapour has to condense itself into little drops of liquid water or ice crystals. This is the first step in the formation of the clouds and precipitation, as well as fog, dew and frost.

What interests us most here is the formation of little drops through the rising air. As vapour condenses, we once again have an exchange of latent heat, this time given off by the vapour to the air around it, which becomes warmer. Here, we have a mechanism of direct heat transport from the surface to the upper troposphere: the water evaporates, absorbing heat (from the sea, for example), the air that contains the vapour is moved upwards by atmospheric motions and it cools as it expands due to the lower pressure. Up high, the vapour condenses into droplets, releasing the heat it has absorbed from the sea, meaning we have a direct heating of the atmosphere at altitude by the sea located thousands of metres below.

In order for the condensation process to take place efficiently, it is necessary for there to be minuscule impurities in the air, such as salt crystals or microscopic mineral granules, soot or even pollen and fragments of phytoplankton. These impurities act as condensation nuclei for the vapour, which gathers around them and covers them in a liquid or frozen film that grows rapidly through condensation, particularly if the impurities are water-soluble. This is how clouds are formed, the clouds that characterize our planet so profoundly and play an essential role in the global climate. For anyone wanting to know more, I recommend the wonderful book *From Raindrops to Volcanoes* by US meteorologist Duncan Blanchard, who provides a simple introduction to the complicated world of interaction between the sea and the atmosphere.

A Cloudy Planet

The droplets and ice crystals formed through the condensation of vapour are the main constituents of the clouds, from the delicate cirrus in the high atmosphere to towering storm clouds. In many varying and constantly changing forms, the clouds have always piqued the imaginations of those observing them. They beg to be admired and are perhaps the most striking characteristic of the Earth when seen from space. At any given moment, around two-thirds of our planet is covered by clouds. There are many different kinds and they are classified according to their characteristics, altitude and formation processes. The first *Cloud Atlas* was published in 1896, and the latest version (from 2017) is available on the World Meteorological Organization's (WMO's) website. What interests us here is distinguishing between convective and stratiform clouds due to the different processes to which they give rise, though we must also remember the high clouds such as cirrus.

The convective clouds form due to an instability in the air column heated from below, as happens in storms. The vertical motions are very intense, the cloud develops predominantly in height and the powerful turbulence keeps the droplets suspended for what can be an extended amount of time. Stratiform clouds, however, tend to spread horizontally and are characterised by weaker vertical motions. Both kinds can give rise to intense precipitation, with convective precipitation generally localized and stratiform precipitation more widespread and homogeneous. Last, but not least, are the cirrus, the wispy clouds of the high atmosphere that form at altitudes between 5,000 and more than 13,000 metres, which are rarefied and predominantly made up of ice crystals. These do not cause precipitation but they do play an important role in the climate.

In the equatorial regions on the boundary between the two Hadley cells, where the air sometimes rises violently due to a powerful warming of the surface, cloud cover and precipitation are extremely abundant and have generated a strip of equatorial forests on all continents. Conversely, clouds tend to be scarce in the tropical regions where we find the descending branches of those same cells. Here, the air that moves downwards is compressed and warmed, and the tiny water drops evaporate, becoming vapour once more.

On the margins between the Ferrel cell and the polar circulation, the air rises once more and in these regions we also find persistent cloud cover, as anyone who has spent time in Scotland or in other areas with latitudes around 60° north or south are well aware. At mid-latitudes, however, the rise and fall of the air is associated with the sequences of cyclones and anticyclones that arrive from the west. Characterized by converging air that rises, extratropical cyclones with their team of atmospheric fronts are associated with intense cloudiness and abundant precipitation. Anticyclones, on the other hand, are characterized by descending air, an absence of clouds and conditions of relatively low humidity.

At these mid-latitudes, and even more so in tropical regions, spectacular storm cells can form when the surface of the planet or the sea warms intensely, generating significant instability in the air column and unleashing violent vertical motions with clouds that rise up to the tropopause. At that altitude, the stable stratification of the stratosphere slows and stops the ascending currents, and the clouds take on their characteristic anvil-like form, commonly seen over warm seas in the summer. Individual storms can sometimes group together in large systems of convective cells or in long, almost linear rows (known as squall lines) of powerful storms that move in an organized way, generating torrential rain and severe damage to people and things.

Even more violent are the hurricanes, which we referred to in chapter 8, in which the rising air movements, forced by the heat given off by a very warm sea, combine with the rotation of the entire convective system leading to a spectacular, swirling structure with a diameter hundreds of kilometres wide and surface winds that can surpass 250 km/h. At their centre is the famous 'eye of the storm', free of clouds and with only weak winds, while all around it a wild chaos of clouds, winds and torrential rains is unleashed.

Hurricanes gain their strength from the heat of the sea and when they encounter a continent, they lose energy and are slowly extinguished, though not without leaving ample destruction in their wake. Because they are driven by a very warm sea with surface temperatures above 26 °C and their formation process is more efficient at lower latitudes, hurricanes are a phenomenon typical of tropical regions, though once formed they can move towards mid-latitudes, as sometimes happens along the Atlantic coast of the United States, for example. Naturally, in a hotter world, one in which the sea temperature was significantly higher, convective structures similar to hurricanes could form at higher latitudes, as it is believed happened during the Eocene and as could happen as a result of global warming.

Something similar is already happening in the Mediterranean, where satellites have revealed the formation of 'Medicanes' (Mediterranean hurricanes), which form at the end of the summer on the warm waters of the Gulf of Sidra before moving towards the coasts of Sicily and Calabria. These meteorological phenomena, fascinating yet dangerous, are associated with intense precipitation and have a formation mechanism that is a little different from that of hurricanes. In general, they form as 'normal' extratropical cyclones, which then, moving over the warm surface of the Southern Mediterranean at the end of the summer, absorb energy from the sea that makes them stronger and causes them to take on the structure of a small hurricane that even has a cloud-free eye and a strong rotation of surrounding air. These localized events tell us how, in a hotter world, hurricanes, or something very similar, could extend beyond their usual latitudes.

Rain and Snow

Farmers have always struggled against bad weather, be it the lack of rain and drought (as in John Steinbeck's book *To a God Unknown*) or an excess of precipitation, floods, hail and storms that destroy the harvests. The variability of the weather has occupied a central place in our concerns for our own survival ever since we abandoned our nomadic lives as hunter-gatherers, if not before.

Let's look, then, at how precipitation is formed according to what we have learned so far. Everything begins with the condensation of water vapour into droplets, which form the clouds. The condensation process is slow and the cloud droplets can last for days, slowly descending, evaporating when they reach altitudes where the temperature is higher and the air becomes unsaturated, and being substituted by new droplets from the areas where irregular motions inside the clouds cause more humid air to rise and cool, condensing the vapour into water droplets.

But this is not enough to cause precipitation as the droplets are too small. In order to generate rain and snow, the droplets must join together to form drops (or ice crystals) that grow ever larger, and that, due to their weight, can no longer be suspended by the turbulent motions within the clouds. When the process of collision and coalescence begins, everything becomes faster and within a few dozen minutes, the drops that have steadily grown ever larger descend rapidly towards the surface.

Often, the first generation of raindrops evaporate immediately beneath the cloud, especially in high temperatures. On the sea in the summer, it is not uncommon to see rain showers that fail to reach the ground because the drops evaporate as they fall. But this process cools the air (due to the absorption of latent heat during evaporation) and, at a certain point, the drops no longer evaporate but fall towards the ground and, hey presto, we have rain!

We have spoken about drops, which are liquid water, but most clouds are found at altitudes where the temperature is below the freezing point. As such, it is better to consider the condensation and collision of ice crystals, which then fall to the ground. These are called 'cold clouds', clouds in which precipitation initially forms as ice, while 'warm clouds' are those found in areas with very high temperatures (or at low altitudes) where water remains in a liquid state.

As the ice crystals and snowflakes fall towards the ground, they encounter layers with ever greater temperatures. If these are higher than the freezing point, the crystals melt and transform into raindrops. Otherwise, they fall to the ground as snow, maintaining their crystalline structure and painting the planet's surface white.

The formation processes for precipitation are, of course, much more complex than we have discussed here. We haven't discussed hail, for example, which also occurs in the summer due to violent, turbulent motions in convective clouds, or the role of lightning and the many other processes that govern this essential component of Earth's climate. For anyone wanting to further their knowledge, the book *Precipitation: Theory, Measurement and Distribution* by English hydrologist Ian Strangeways provides a complete, technical picture, with historical information on the world of precipitation and its measurement.

The Little Boy off the Coasts of South America

In chapter 7, we saw how currents in the atmosphere and the ocean are closely linked to one another, because the winds move ocean waters while the distribution of marine heat modulates and powers the winds in the atmosphere. Among their various modes of interaction, one of the most important is connected to the juxtaposition between the marine phenomenon known as *El Niño* and the atmospheric Southern Oscillation of the trade winds in the equatorial region of the Pacific. In the 1920s, the Southern Atmospheric Oscillation was described in detail by British climatologist Sir Gilbert Walker, but without highlighting its connection to marine circulation. Instead, it was Norwegian meteorologist Jacob Bjerknes who, in the 1960s, recognized the close connection between these atmospheric and marine oscillations. This phenomenon is now referred to collectively as ENSO, from the initials of the two components, and it influences the precipitation for most of our planet, even in regions located a long way from where ENSO originates.

The name *El Niño* (the Little Boy) comes from the fact that, around Christmas time, with irregular cadence, a particularly intense phenomenon can occur along the Pacific coast of South America. Here, there tends to be a resurfacing of deep, cold waters (upwelling) that are rich in nutrients and attract fish in great numbers. But during an *El Niño* episode (known also as the positive phase of ENSO), the marine waters along the South American coasts warm in an anomalous way, the water's resurgence becomes weaker and fish levels decline rapidly. Not a particularly nice Christmas present for the fishermen of Ecuador and Peru.

This phenomenon is generated by a coupled oscillation in both atmospheric and oceanic circulation. Equatorial winds (the trade winds) generally push the waters from east to west, recalling the deep waters on the Eastern edge of the basin and accumulating the warm surface water on the Western margin. On Asiatic coasts, the air rises and generates intense rainfall. But, because of the Southern Oscillation, the trade winds are periodically weakened, causing less water to resurface along the coasts of South America and moving convective clouds and precipitation towards the centre of the Pacific, drastically changing the hydrological regime throughout the region. This means *El Niño* has arrived.

In order to understand how the oscillation is generated, it is necessary to consider the fluctuations in marine currents that spread towards the east along the equator (through oscillations known as Kelvin waves) and towards the west at tropical latitudes as fluctuations called Rossby waves. An entire oscillation cycle of the coupled atmosphere–ocean system of the equatorial Pacific has an average length of 5 years, though this period is actually very irregular and can last between 3 and 7 years, the typical time observed today between two successive episodes of *El Niño.*

Variations in marine and atmospheric circulation caused by the alternation between the *El Niño* phase and the opposite situation – known, for reasons of symmetry, as *La Niña* (the phase in which the trade winds and the resurgence of water along the coasts of South America are more intense) – influence the climate and levels of rainfall in vast areas of the world, alternately causing drought or extensive flooding. ENSO is a phenomenon of tropical and equatorial origin, but one that is capable of causing an impact on a global scale, characterizing the planet's climate. Figure 9.1 shows the two situations: the 'normal' one (*La Niña*) and that of *El Niño.*

Low Clouds, High Clouds

Let's get back to clouds, which play a central role in the dynamics of the climate due to the significant effects they have on the albedo and the greenhouse effect. Clouds are white and, therefore, reflect solar radiation and increase the albedo, contributing to a cooling of the surface below, which receives less light. At the same time, water vapour and cloud droplets have a significant greenhouse effect, as they absorb infrared radiation and re-radiate it towards the surface, warming it. These are two opposite effects, which make the cloud's impact somewhat delicate and complicated to estimate.

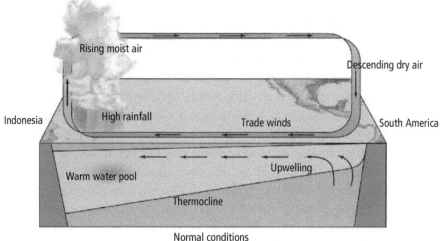

Rising moist air

Descending dry air

Indonesia

High rainfall

Trade winds

South America

Warm water pool

Upwelling

Thermocline

Normal conditions

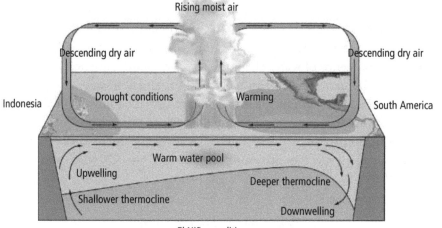

Rising moist air

Descending dry air

Descending dry air

Indonesia

Drought conditions

Warming

South America

Warm water pool

Upwelling

Deeper thermocline

Shallower thermocline

Downwelling

El Niño conditions

Figure 9.1 The normal condition of tropical circulation in the Pacific (above) and that of *El Niño*, characterized by warm waters and scarce upwelling of deep waters close to American coasts and a shifting of precipitation towards the central zone of the ocean. Taken from the European Space Agency (ESA): www.esa.int/ ESA_Multimedia/Images/2018/08/El_Nino.

Today, clouds tend to have a cooling effect on the Earth's climate, reflecting around 50 W/m^2 of solar radiation and re-radiating around 30 W/m^2 downwards. The difference is around 20 W/m^2, which leads to a cooling of around 5 °C of the Earth's surface. So, if there were no clouds, the average temperature would be significantly higher.

But not all clouds are equal. With cirrus clouds, for example, the greenhouse effect is greater than their reflective abilities, while for lower clouds, whether convective or stratiform, the albedo effect dominates. The overall effect depends, therefore, both on which clouds are most present and the reflective capacity of low clouds. When there is pollution by dust or soot (carbon compounds produced by combustion, collectively called black carbon), clouds lose part of their reflective capacity and therefore cool less.

This brings us to the extremely difficult question of what the distribution of the clouds might look like in different climate situations. What happens if the atmosphere's average temperature rises by 2 °C? Or by 4 °C? What if there is an increase in soot emissions? These questions are at the heart of current climate research, because behaviour of the clouds is one of the greatest uncertainties, which must be eliminated in order to carry out reliable climate projections for conditions different from those we have today. If, as various recent studies suggest, clouds in a warmer world could lose part of their ability to cool or even become less abundant, then the balance between the greenhouse effect and the albedo could change and the response by the clouds could amplify the warming caused by the rise in the concentration of carbon dioxide in the atmosphere.

Clausius and Clapeyron Again!

The clouds' response is not the only way the water cycle can amplify global climate changes. At the beginning of the chapter, we saw that the quantity of vapour that can be held in the atmosphere before it condenses as droplets depends on the air's temperature, according to the Clausius–Clapeyron law.

If the temperature drops – for example, because the concentration of carbon dioxide or methane falls – then much less vapour will remain in the atmosphere as the colder air will become saturated by much lower quantities of vapour. Therefore, part of the vapour present in the atmosphere has to be condensed as precipitation and finally fall to the ground. But water vapour has a powerful greenhouse effect and any decrease amplifies the initial perturbation in temperature. The world becomes colder, the air less rich in vapour and the entire evaporation and precipitation cycle slows and becomes weaker. A frozen planet, like that during the Snowball or a glaciation, is a dry world with strong winds and little precipitation. A little like we see today in Antarctica.

Conversely, a rise in carbon dioxide in the atmosphere would cause an increase in the air temperature, which could then host a larger amount of water vapour. There are two possibilities here: that the quantity of water vapour remains the same, causing the relative humidity to decrease (defined above as quantity of vapour actually present divided by the maximum quantity that there could be), or that the relative humidity remains fairly constant, therefore recalling vapour through evaporation or transpiration from terrestrial surfaces. The data and observations tend to suggest that this second possibility is generally the most likely, as we can see from the fact that on Earth there is no shortage of evaporation sources, given that oceans cover 70 per cent of the planet's surface.

So, here we have another powerful amplification mechanism: the hotter it is, the more vapour can be held in the atmosphere and the greater the greenhouse effect. At the same time, more vapour means more energy for atmospheric motions, meteorological perturbations that can be stronger, rainfall that can become much more violent. The climate of the Eocene was probably like this, with an unbridled water cycle and hurricanes not confined to the tropics. But even a rise of just a few degrees in temperature can modify the precipitation regime and lead to a greater frequency and intensity of extreme meteorological events.

The Water Cycle

Water evaporates from the sea and the lakes and is transpired by plants, transforming into vapour and absorbing heat from the surface. It then rises into the atmosphere where it contributes to the greenhouse effect. Here, it condenses into droplets, forms clouds and is transported from

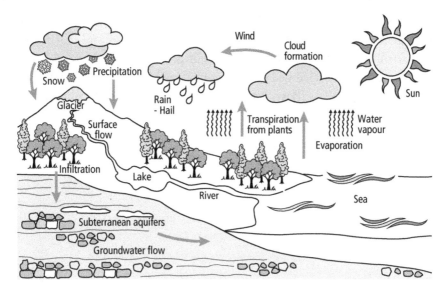

Figure 9.2 A simplified diagram of the Earth's water cycle with its various components.

one side of the Earth to the other by the winds until the processes of coalescence begin that generate precipitation as rain, hail or snow. This falls back onto the surface of the planet, sometimes very close to where the water had evaporated, other times thousands of kilometres away.

The water runs on or below the surface, forming aquifers that are rich in mineral salts ripped from rocks during the processes of erosion and weathering, sometimes unfortunately contaminated by all sorts of pollutants. It flows and, carrying with it a message from the continents – be it good or bad – pours into that great tank that is the global ocean. Some of this water is intercepted by plants, absorbed by the roots and transpired before it reaches the sea, before then returning to the atmosphere on a shortened journey.

The snow that falls in winter lies in wait for many months before melting in the spring and returning to its planetary path. When it falls high in the mountains or at the poles, the snow can instead solidify slowly into ice, forming mountain glaciers and polar ice sheets which also flow towards melting, albeit much more slowly and majestically.

The water cycle is perhaps that which tells us most immediately about the circularity of nature, the continuous transformations and that *Pánta rheî* attributed to the philosopher Heraclitus. Everything flows with the water cycle that is changed by the climate, which in turn it helps to determine. Figure 9.2 shows a simplified diagram of this global cycle.

10

The Planet Cools

The crocodile said, "Where did all those forests and swamps that covered the Earth all the way to the poles go? Why is everything colder, drier and dustier now, with the ice ever advancing? And what about those strange creatures that now run through the newly-formed prairies, eating grass, and those others, even faster, that follow, trying to catch them? What kind of a new world is this?"

The Climate's Pendulum

After its culmination in the hot period of around 50 million years ago, the climate's slow pendulum swung back the other way and the planet began to cool, sometimes slowly, sometimes suddenly, sometimes even with temporary returns to warmer conditions. The Zachos curve (shown in chapter 8) reveals how global temperatures dropped by around 10 degrees between the Eocene and today, during the geological era known as the Cenozoic, which began around 65 million years ago and is still ongoing. This is the era of the mammals that expanded throughout all environments after the disappearance of the non-avian dinosaurs, exterminated by an asteroid and volcanic eruptions.

The slow oscillations of the climate should no longer surprise us, but our planet's 'recent' history is a mine of information with which to understand how the Earth System works, thanks to the wealth of paleoclimatic data provided by sediment coring and isotopic analysis. In particular, the measurements show that, during the Cenozoic, the concentration of carbon dioxide in the atmosphere had already decreased some few millions of years ago from the incredibly high levels found during the Eocene to the typical concentration of the pre-industrial period at less than 300 ppm.

Parallel to this decrease in temperature, glaciers began to form at the highest altitudes and in polar regions, reflecting solar radiation and trapping ever-greater quantities of water in a solid state. Sea levels dropped, the climate gradually became less rainy and all of the ecosystems moved towards great change, creating landscapes increasingly similar to those we can see today.

The forests, present almost everywhere in the warmer and rainier climate of the Eocene, were gradually substituted in many areas by prairies and expanses of grass. One of the reasons for this change might have been an increase in wildfires, facilitated by the still elevated temperatures and the decreasing rainfall. The fires presumably had a more destructive impact on the large trees, which are more vulnerable to fire, than on the grass in the prairies, which had a competitive advantage. In the Cenozoic, the savannahs that today cover vast regions of our planet began to expand, perhaps providing a starting point for the long human adventure. The deserts did the same thing.

In response to these changes in vegetation, animals also evolved, adapting to the new open spaces. Among the herbivores, for example, 'browsers' such as goats and ibex, which fed predominantly on leaves, shoots and fruits from trees and bushes, declined, while the population of 'grazers', such as sheep, bison and horses, which live on grass and prefer open spaces, grew. At this time, the evolutionary history of the horse unfolds, starting with the small *Eohippus* of the Eocene, moving through the usual complicated 'evolutionary tree', and ending with their selection by humans, leading to the large dray and saddle horses whose use characterized our civilisation until a few decades ago. Predators also adapted, evolving a greater capacity to hide among the long grass and suddenly pounce on some unlucky herbivore in a deadly competition of who runs fastest and who can see danger or prey the best in the new environments of the ever expanding prairies.

An Antarctic Story

During the long period of cooling that began in the Eocene, there were a number of particularly significant episodes associated to faster changes and sudden decreases in temperatures. The curve of oxygen-18 around 34 million years ago, between the end of the Eocene and the beginning of the Oligocene, reveals a significant decrease in global temperature of around 3–4 °C, and a robust increase in continental ice.

What essentially happened at that time (which lasted around 400,000 years) is the stabilization of the Antarctic ice sheet, which had begun to form around 10 million years earlier. Antarctica had started to accumulate increasing amounts of ice, transforming itself from a continent almost entirely devoid of ice, as it was in the Eocene, into an expanse of mountain glaciers and then a real ice sheet, even if it was still much smaller than its current dimension. The East Antarctic ice sheet, today the largest and up to 4,800 metres thick, was the first to form.

The decrease in global temperatures during the transition from the Eocene to the Oligocene was further amplified by the high albedo of the ice that was forming and that further reduced the quantity of solar energy absorbed. It also contributed to the heightened geographical isolation of the Antarctic continent, which found itself surrounded by the Antarctic Circumpolar Current and a strong circumpolar vortex in the atmosphere, which inhibited the influx of warmer air or water from the southern mid-latitudes. And so, the second Antarctic ice sheet was formed: the West Antarctic ice sheet.

Later, some 13.8 million years ago, there was a second episode of intense cooling, which led the two Antarctic ice sheets to take on more or less their current dimensions. Since then, despite significant oscillations, the Antarctic has remained covered in ice, especially on its eastern sheet, while the western sheet has undergone larger variations and continues to be the most unstable and at potential risk of collapse. In this period, glaciers were certainly present in the Northern hemisphere but there probably wasn't an actual ice sheet there, not even in Greenland, where it would later begin to form.

The detailed analysis of fluctuations in temperature and the volume of the ice around the transition of 13.8 million years ago, as ascertained from the isotopic ratios between oxygen-18 and oxygen-16, reveal the existence of 'rapid' oscillations over a period of around 40,000 years before the transition and 100,000 after the decrease in temperature. These oscillations are very important because they show us the planet's climate response to variations in the characteristics of the Earth's orbit, in particular to its obliquity (the slope of the Earth's axis) and its eccentricity. But it would only be later, in the last 3 million years, that orbital changes would become a dominant factor in the Earth's climate.

Is It the Himalayas' Fault?

The fundamental question is, of course: what caused the cooling of the last 50 million years? And the most probable answer once again involves the decrease of carbon dioxide in the atmosphere, which declined dramatically during the Cenozoic. Other more or less picaresque options – sometimes suggested as 'definitive' explanations – have been disproved by the data. Naturally, the reduction in the greenhouse effect did not act alone, unleashing the now well-known cascade of amplification mechanisms, ranging from the reflection of solar radiation by the enormous frozen blanket that was being formed to the changes in marine and atmospheric circulation and the decrease in water vapour contained in the atmosphere.

So, then, the real question is: what did cause such a significant decrease in atmospheric carbon dioxide? And this is where the scientific discussion gets interesting. Over periods lasting millions of years, the concentration of carbon dioxide is heavily controlled predominantly by the geological carbon cycle and, in particular, the balance between the processes that remove CO_2 through surface rock weathering and those of volcanic emissions. It is necessary, therefore, to understand which of these mechanisms could have been responsible for the change in the concentration of carbon dioxide.

The most important geological event of the Cenozoic was the collision of the Indian plate with the Asian continent, which led to the formation of the long Himalayan mountain range. In 1988, US marine biologist and paleoclimatologist Maureen Raymo and her team suggested that this episode profoundly influenced the concentration of atmospheric carbon dioxide and, therefore, the planet's climate.

The collision between India and the Asian continent began between 50 and 40 million years ago and continues today. The collision of these two continental plates gave rise to the highest mountain range in the world today and formed the enormous Tibetan plateau. The presence of a large mountainous region at such altitudes strengthened the Indian monsoon and drastically changed the atmospheric circulation of the entire Northern hemisphere, leading to greater seasonality and changes in the geographical distribution of precipitation. But this change on its own is not enough to explain the drop in carbon dioxide.

Something else was needed. Such as a rise in the weathering of the rocks that emerged during the formation of Tibet, with a corresponding acceleration in the absorption of carbon dioxide. On the basis of previous studies, Raymo and her co-authors suggested that the speed of rock weathering depended not only on the exposed surface, but also – if not more so – on the relief features of the continents. And it is difficult to find a relief as imposing as that of the Himalayas.

So, we have a possible culprit for Cenozoic cooling: the decrease in carbon dioxide caused by greater efficiency in the processes of erosion and chemical weathering of surface rocks and their absorption of CO_2. If this view is correct, the birth of the largest mountain range on Earth caused the sequence of events that led the planet towards a cooler, drier climate. In addition to this, more recent work by a group of paleo-climatologists and scholars of magmatic processes – comprised of Pietro Sternai, Luca Caricchi, Claudia Pasquero and their team – demonstrates how, during the Cenozoic, magmatic activity also diminished, with a consequent reduction in carbon dioxide emissions and, therefore, a strengthening of 'Himalayan cooling'. As often happens with the climate, different contributing factors could have once again acted simultaneously to bring about the most significant changes of the Earth System.

Old Trees and New Grasses

The cooling of the Cenozoic brought important changes to the ecosystems and vegetation. Before the decrease in temperature, the climate was hot and humid, the Earth was covered in forests (even the polar continents) and there were very few deserts.

The trees have a particular way of fixing carbon through photosynthesis, known as the C3 cycle. In this metabolic process, the enzyme RuBisCO (a nickname that stands for the much longer ribulose-1,5-bisphosphate carboxylase/oxygenase, the main protagonist in modern photosynthesis) transforms carbon dioxide into organic compounds with three carbon atoms (hence C3). However, due to imperfections in RuBisCO, around 30 per cent of the fixed carbon is lost and oxidized once more into CO_2, in a process called photorespiration. A forest environment is essentially a world dominated by C3 metabolism.

However, another metabolic cycle, C4, is possible. In this process, a CO_2 molecule is attached to an organic compound with three carbon atoms (hence the name C4). These are then all transported close to the RuBisCO, where the carbon dioxide molecule is released, generating an environment around the enzyme with a high concentration of CO_2 and suppressing photorespiration. Today, many herbaceous plants in tropical and temperate zones essentially use the C4 cycle, which is favoured by conditions of scarce precipitation and high luminosity.

Both of these metabolic processes have advantages and disadvantages, and different energetic costs. The higher the atmospheric concentration of CO_2, the easier it is for the C3 cycle because photorespiration slows down. During the Eocene, atmospheric concentration of CO_2 was much higher, the world was rainy and, overall, trees with their C3 photosynthesis were favoured.

With the decrease in carbon dioxide concentration and greater environmental aridity, the herbaceous plants (typically C4) became more competitive. As we can see from analysis carried out by US geologist Thure Cerling and his colleagues, as of 10 million years ago C4 plants continued to expand, and today around 75 per cent of the plants are C3 and 25 per cent are C4.

But how did we discover that the plants have changed? The answer lies in the teeth of herbivores! The starting point is that C3 and C4 plants assimilate carbon-12 and carbon-13 in different ways (we discussed isotopic ratios in chapter 5) and so their tissue has a different ratio to the abundance of the two isotopes. These differences are preserved in the tissues of herbivores which feed on plants of one kind or another, and in their teeth in particular, which are more easily preserved over millions of years. And it is by analysing the carbon isotopic ratios in a large number of fossilised herbivore teeth that Cerling's research group was able to prove the growth in the abundance of C4 plants starting 10 million years ago.

The complex story of the collision between two plates, the formation of the Himalayas, the subsequent change in climate and the response from the plants, ecosystems and herbivores shows once more how, on our planet, everything is connected and belongs to those global planetary cycles that we have already looked at and that we will discuss again.

Where Did All the Water Go?

The Mediterranean, contained between Europe in the north and Africa to the south, is a sea with a water deficit, as the evaporation caused by high temperatures and strong insolation is larger than the water provided by precipitation and rivers. As a consequence, the Mediterranean is saltier than the Atlantic, to which it is connected today by the Strait of Gibraltar, through which the superficial Atlantic water flows into the Mediterranean like a torrent descending to fill a lake. Deep down, however, the stream flows in the opposite direction, since the greater salinity and density of the water from the Mediterranean generates a counter-current that crosses the strait, taking saltier water to the heart of the Atlantic.

If for some reason the Strait of Gibraltar were to close, this water exchange would cease and the Mediterranean would slowly evaporate, leaving only briny lakes on what was once the seabed. This apocalyptic scenario effectively took place around 6 million years ago and lasted for almost 700,000 years. At the time, the area around Gibraltar was much more open than it is today but also much shallower. Due to movements in the continental plates, the processes of tectonic uplift and a decrease in sea level due to a rise in glaciers, the opening at Gibraltar temporarily closed, causing the Mediterranean to dry up.

The existence of this kind of episode, which happened during the so-called Messinian stage, was definitively proven in the 1970s by the measurements taken by US oceanographic vessel *Glomar Challenger* during the Deep Sea Drilling Program. Italian geologist Bianca Maria Cita played a central role in this discovery, together with a sizeable team from all over the world and coordinated by Japanese-American geologist Kenneth Hsu, based at the time in Switzerland.

During that period of time, the entire Mediterranean evaporated, leaving an arid depression in some areas that was more than 3 kilometres below the level of the Atlantic, which foamed on the other side of Gibraltar. On the seabed, vast salt deposits formed that are still visible today in the drillholes and in various outcrops in Sicily and the Italian Apennines. Sometimes, water would return from the Atlantic before evaporating once more in a sequence of events that definitively ended around 5.3 million years ago, when the gap that would become the

Gibraltar Strait began to open. Imagining the waters of the Atlantic as they pour in, perhaps first as a trickle before becoming an increasingly powerful waterfall filling the Mediterranean seems more far-fetched than a disaster movie. But it really happened.

Before this 'grand finale' that was so portentous, during the Messinian the great difference in height between the Alps and the arid bed of the Mediterranean basin caused a reactivation of all rivers, which forged spectacular canyons on the boundaries between the mountains and the salty plains thousands of metres below. This is also how the deep incisions were made that would later lead to the formation of the great Italian pre-Alpine lakes (Maggiore, Como, d'Iseo, Garda), and which are still visible today through seismic prospecting in the layers of rock and sediment beneath the lakes. Later, the actions of the glaciers over the last 3 million years shaped the landscape and the relief, creating the soft forms we see today, but beneath lie the canyons of that strange, salty, arid landscape.

Movements inside the Earth, therefore, have a crucial effect on the Earth's climate and environment. From the formation of the Himalayas to the closure of the Gibraltar strait, the upheaval caused by the Earth's crust and mantle inevitably influences what happens on the surface. It is therefore always necessary to keep them at the front of our minds, avoiding any insistence on being able to comprehend the Earth System without understanding the dynamics of its slowest and most powerful component.

11

The Breath of the Ice

They looked at the cold plain, arid and windswept. A few animals grazed in the distance, but it was so difficult for them to find food, to hunt for large and small herbivores and feed their children. And yet, deep down in the caves at the heart of the Earth, they had been creating masterpieces for generations, masterpieces that would survive for millennia and that depicted ibex, bison, the immense aurochs and all the other animals in order to pacify their spirit, to pray that they leave their gifts and to live in harmony with the Great Mother.

Discovering the Glaciations

The idea that in the past there were periods with much more extensive ice coverage than we have today has long been part of our culture. But this was not always the case. Before the Arctic and Antarctic explorations, it was difficult for scientists to imagine enormous glaciers, some thousands of metres thick. The experience of the Alps, the Andes and the Himalayas showed the existence of great 'rivers of ice', though always confined to valleys at high altitudes.

It was in the first half of the nineteenth century, when observing the broad distribution of erratic masses – those great blocks of rock shifted by the movement of glaciers that had long since disappeared and that even can be found some distance from the glaciers themselves – that various scientists sensed the role played by the glaciers as forces capable of profoundly shaping the landscape. German scientist, writer and artist Johann Wolfgang von Goethe, Swiss geologists and alpinists Horace-Bénédict de Saussure and Jean de Charpentier, engineer and glaciologist Ignaz Venetz, and scientist and poet Karl Friedrich Schimper were the first to grasp the relevance of glaciers as primary geomorphological forces, and to recognize that there could have been at least one moment in the past when they had been much more extensive. In 1833, Venetz published a book in which he suggested Europe had previously been covered almost entirely by ice, formalizing the idea of an ice age.

After having studied and analysed a great deal of observed evidence, in 1837 Swiss paleontologist and glaciologist Louis Agassiz set forth his theory that the Earth had gone through at least one significant ice age, and in 1840 he published a two-volume work on glaciers that consolidated most of the information available at the time. In these volumes he discusses, among other things, the formation of moraines and the so-called 'roche moutonnée' (literally, sheepback rocks, but better known as 'whaleback-shaped rocks' in the English-speaking world), carved by the movement of the ice and grit that glaciers bring with them. Shortly after, Agassiz moved to the United States, staying there to work until the end of his life and becoming one of the most famous scientists of his time. Unfortunately, he also assumed positions that were anti-Darwinian and creationist while also developing biological theories that were openly racist.

But people struggled to accept the idea of a Europe covered in a blanket of thick ice, particularly due to a lack of direct experience with such a scenario. The exploration of the Arctic, however, was just around the corner, and the first accounts from the ships returning to Europe from the North Pole told of colossal walls of ice hundreds of metres high, gigantic frozen blocks that fell into the water and floated for weeks, an unknown world inhabited by people who had adapted to life in the great cold – all of which seemed to be the realization of the ice age imagined by Charpentier, Venetz, Schimper and Agassiz. The idea of an ice age finally became acceptable and, since then, discoveries have followed one another at an increasingly impressive rate.

Children of the Ice Age

A little less than 3 million years ago, a long time after the Mediterranean was back to being a real sea, our planet fully entered its current ice age, a time of almost regularly fluctuating climate – the so-called sawtooth climate pattern characterized by the succession of long periods of ice interspersed with brief interglacial periods that were warmer and rainy. This was the Pleistocene, the age of modern glaciations and the evolution of hominids. Figure 11.1 shows the oscillations in temperature

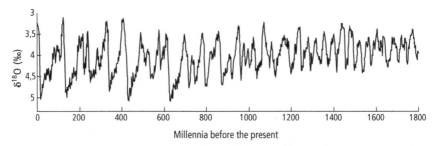

Figure 11.1 Oscillations in temperature and ice extension reconstructed through the isotopic ratio of oxygen-18 to oxygen-16, combining various drillings of deep ocean sediment. Warm periods with little ice are at the top and correspond to low $^{18}O/^{16}O$ ratios, and cold periods with lots of ice are at the bottom. Time runs from right to left and the horizontal axis indicates the thousands of years before the present day. Taken from Lorraine E. Lisiecki and Maureen E. Raymo, 'A Pliocene–Pleistocene stack of 57 globally distributed benthic $\delta^{18}O$ records'.

over the last 2 million years, reconstructed (as always) by measuring the isotopic ratios of oxygen in marine sediment. This is, effectively, the climate in which our ancestors began their evolutionary journey to become, around 300,000 to 200,000 years ago, the species that pronounced itself *sapiens*.

The analysis of isotopic fluctuations reveals the existence of two distinct regimes. Up until around a million years ago, the oscillations between the glacial and interglacial periods were more modest and irregular, with a period of around 41,000 years between one warm phase and the next. These fluctuations showed relative symmetry between the periods of ice accumulation and those of melting.

More recently, these fluctuations increased, the global temperature dropped further, the periodicity became around 100,000 years and we can see a clear symmetry between a slow phase of ice growth, which can last up to 70,000 or 80,000 years, and a phase of more rapid deglaciation and warming typically lasting some 10,000 years. Following the increase in temperature, there was a phase (known as interglacial) with a temperate, more humid climate that ushered in a new, slow growth of the ice. Today, we are living in one of these interglacial periods, though it is not a given that the oscillation will continue as it has up until now. This is because *Homo sapiens* (perhaps not so wise, after all) have significantly disturbed the climate of the Earth and we do not yet know what will happen in the next few hundred years. But we will get to this shortly.

The closer we get to the present day, the more abundant climate archives we have and the more detailed information we can obtain on the variability of the climate over recent millennia. Fossilized pollen conserved in the sediments of lakes or the seabed, for example, tell us what kinds of plants were present in a particular area or time period. On the basis of what we know about the environmental needs of vegetation, we can reconstruct the climate of that time. Similarly, speleothems (calcareous deposits found in caves, such as stalagmites) are formed by the precipitation of successive mineral layers from subterranean water, and their isotopic composition preserves traces of the climate conditions at the time of their formation.

But it is drilling (or coring) in the Antarctic ice that has, more than anything else, provided us with an incredible amount of information on the climate of the last million years. Deposited at a very slow rate due to scarce precipitation and extremely low temperatures, and having been barely modified since, Antarctic ice preserves characteristics from the Earth's climate hundreds of thousands of years ago. The ratios between stable oxygen and hydrogen isotopes allow us to estimate the temperature at the time the ice was formed, while the dust content provides information on the aridity of the climate. Finally, air bubbles trapped, sealed and tightly enclosed in the ice contain the atmosphere of that ancient world and allow us to measure directly the carbon dioxide concentration at the time of the ice's solidification.

In 1998, an expedition on the ice sheet by an international group of Russian, French and US scientists around the Russian Antarctic station of Vostok led to the extraction of a thin cylinder of ice, known as an 'ice core', that revealed the climate history of the last 400,000 years. In 2004, the European Project for Ice Coring in Antarctica, EPICA, extracted an even deeper ice core that registered fluctuations in temperature, dusts, carbon dioxide, methane and other environmental parameters over the last 800,000 years. In 2008, a working group involving a large number of European paleoclimatologists published a fundamental work in which they described the initial results of the coring and the data analysis, which was later refined in 2015 for the older period.

Figure 11.2 shows the temperature curve reconstructed from the isotopic ratio of deuterium to hydrogen and the concentration of carbon dioxide contained in the gas bubbles, all obtained from data gathered by the EPICA project. We can see the slow accumulation of ice associated with the decrease in temperature followed by a rapid deglaciation and a warmer interglacial period, with an approximate repetition of the entire cycle, on average, every 100,000 years. The difference in temperature between a glacial and an interglacial maximum, estimated using isotopic ratios, can be greater than 6 °C and it is important to note how, in the last 800,000 years, the climate system has varied between extremes that were always very similar. At least up until around 200 years ago.

Greenhouse Gases and Glaciations

Figure 11.2 shows us that the temperature record (obtained through isotopic analysis) varies in step with that of carbon dioxide. A similar thing happens for the concentration of the methane obtained from air

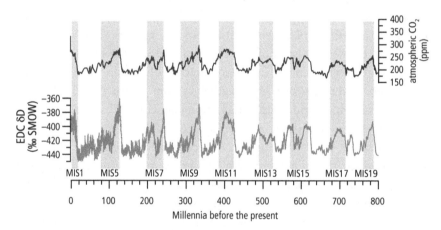

Figure 11.2 Record of the isotopic ratio of deuterium to hydrogen (bottom curve) that provides a temperature estimate (growing towards the top), and the content of carbon dioxide in ppm (top curve) measured in air bubbles trapped in the ice. The horizontal axis shows the thousands of years before the present. The vertical grey bands show the sequence of marine isotopic stages (MIS) used to identify the different paleoclimatic periods. Redesigned by Eleonora Regattieri using data from EPICA.

bubbles. But which one comes first? This is important as it allows us to reveal the links between variations in greenhouse gases and temperature.

The problem is not a simple one, as both the isotopes and the water bubbles move slightly in the ice even after they are deposited (or trapped), but they do not do so in the same way. It is therefore necessary to carry out an extremely precise dating, bearing in mind the possible movements that can happen after the ice is formed. This work has continued over the years since the EPICA ice coring and has reached a significant conclusion: namely, that fluctuations in global temperature generally follow a fluctuation in carbon dioxide with a variable time delay, though such a lead–lag relationship is sometimes opposite, depending on the time period considered.

In order to understand better how these delays are created, we need to remember that the ocean contains a much higher quantity of CO_2 than the atmosphere, and variations in marine circulation (for example, thermohaline flows in the deep ocean) can have an effect on the atmosphere. Furthermore, the solubility of carbon dioxide in water depends on the temperature, in such a way that the colder the water, the greater the amount of dissolved CO_2 it can contain, and vice versa. It is also necessary to verify which temperature we are talking about: a global average or the temperature of the region in which the ice that would later be cored was formed? The concentration of carbon dioxide in the atmosphere is very uniform, but neither the temperature of the Earth's surface nor its fluctuations are equally homogeneous.

As always, the issue is more complicated than it might seem at first glance, because there are multiple interactions between the different components of the climate system, as we have seen on many other occasions. These components are able to self-regulate and amplify or limit themselves in turn, forming Alfred Wallace's 'grand organic whole'. In this case, too, an initial modest increase in carbon dioxide could have led to temperature rises that in turn could have caused the permafrost to melt and emit more greenhouse gases, or it could have caused more carbon dioxide to be released from the ocean that had grown warmer, and more generally, it could have triggered various other amplification mechanisms leading to the melting of the ice shelves. Or, vice versa, an initial decrease in CO_2 could have been amplified by the same mechanisms acting in reverse, leading to an accumulation of ice.

It is also possible that a variation in temperature caused by other factors, such as a reorganization of the thermohaline circulation, changes in the water cycle and in the quantities of atmospheric dusts or changes in the distribution of solar radiation reaching the Earth could have triggered a response by the carbon cycle that altered the concentration of greenhouse gases, reinforcing the initial fluctuation. In short, it is a complex system that should be tackled by considering the functioning of all the various processes at play without preconceived ideas. We must recognize that a small perturbation can be reabsorbed or amplified, generating a new climate and often in unexpected ways. In a 2022 work, for example, a group of theoretical physicists and climatologists led by Marco Baldovin, now at the University of Paris-Saclay, and including myself reconsidered the EPICA data using an approach based on causation analysis, identifying a strong mutual causal influence between temperature and CO_2 concentration fluctuations on millennial time scales.

The important thing is to understand the underlying causes and, perhaps even more so, to evaluate carefully how the climate system responds to these forces. It is a fascinating problem that also has profound implications for our capacity to evaluate the climate's response to the perturbations we are setting in motion today. In the case of glaciations over the last 3 million years, the regularity of the oscillations leads to further questions. Are these oscillations generated within the climate system, or is there an external clock that modulates the ice's breathing? In order to answer these questions, we must push past our planet and explore variations in the Earth's orbit.

Planets in Chaos

When a single, perfectly spherical planet turns around its star, it follows an elliptical orbit with the star situated in one focus of this ellipse. This is Kepler's First Law, formulated by the German astronomer in 1609. The planet regularly moves on the same unchanging orbit, going a little faster when it is close to the star and a bit slower when it is farther away.

A planet's orbit is characterized by various parameters. What interests us here is the eccentricity (or how 'elongated' the orbit is compared to a circumference, which has zero eccentricity) and the obliquity, that is, the inclination between the planet's axis of rotation and its axis of revolution (or the axis that is perpendicular to the plane of orbit). Also important is the direction of the axis of rotation, which can at any given moment in the orbit (for example, at the Spring equinox) 'point' either at the star or in the opposite direction; it can be in a plane that is perpendicular to the segment that unites planet and star or point in any other direction.

When there are lots of different planets, or planets that are not perfectly spherical, or those with large satellites like the Moon to the Earth, the gravitational pull between the different bodies generates disturbances in the movement of all planets, whose orbits are not unchanging over time. Rather, these orbits begin to vary and become complicated, with the positions of planets becoming unpredictable over long periods of time, from hundreds of millions to billions of years, changing to such a degree that the planets can even end up crashing into one another.

This problem, already discerned and described by the great French mathematician Henri Poincaré at the end of the nineteenth century, was precisely formulated and simulated on a computer by French astronomer Jacques Laskar, in 1989 and again over the following years. In particular, Laskar showed how the solar system is characterized by chaotic orbits (albeit over very long periods of time), which make the position of the planets over billions of years unpredictable.

For shorter periods of time, millions or tens of millions of years, the movement of the planets continues to be predictable, but orbital parameters do gradually vary. As such, the eccentricity, obliquity and position of the equinoxes vary over time and can be calculated. But it is precisely these characteristics of the orbit that have a significant influence over the quantity and geographical distribution of solar radiation reaching the planet. The Earth is no exception. Our orbit is slowly changing, and the latitudinal and seasonal distribution of solar energy received varies as a consequence.

Milankovitch Cycles

In 1842, French mathematician Joseph-Alphonse Adhémar hypothesized that terrestrial glaciations were generated by variations in the astronomical parameters of the Earth's orbit. This idea was revisited by Scottish climatologist James Croll in the second half of the nineteenth century, but it was only in 1913 that Serbian mathematician and engineer Milutin Milankovitch published a quantitative study on the variation of insolation following changes of eccentricity, inclination of the axis of terrestrial rotation and precession, and the possible effects these variations could have on the climate.

Milankovitch's study observes, first and foremost, that the orbital parameters have varied in a fairly periodic way over the last few million years. In particular, changes in eccentricity have a main period of around 413,000 years, modified by two additional periods of 95,000 and 125,000 years, which together generate a periodicity of around 100,000 years. The obliquity, however, varies between 21.5° and 24.5° with a period of around 41,000 years, while the direction of the axis of rotation (precession) varies with an approximate period of 21,000 years.

The coincidence between the orbital variations (with a period of 100,000 and 41,000 years) and glacial oscillations is remarkable, and this led Milankovitch to the hypothesis that the latter are generated by these changes in the Earth's orbit. The Serbian mathematician focused primarily on incident solar radiation and, subsequently, temperature during summer in the Northern hemisphere. Several decades later, the analysis carried out on oceanic cores by US geologists and oceanographers Jim Hays, John Imbrie and Nicholas Shackleton allowed for an accurate analysis of climate variations over the last 450,000 years, leading to the modern formulation of the theory for orbital control over glacial oscillations in the Pleistocene.

The mechanism, later refined by researchers such as Maureen Raymo (whom we have already met), US climatologist Peter Huybers and many others, should function more or less like this: the greater the inclination of the Earth's axis, the greater the quantity of energy per square metre reaching the high latitudes (southern and northern) during the summer of that hemisphere, and, accordingly, the less incident energy arriving during the winter. The rise in summer heat should lead to a melting of ice and snow, impeding its accumulation from one year to the next and stimulating deglaciation processes. In contrast, when the obliquity is low, the lower summer temperatures can allow for the slow development of ice sheets in both the Northern and Southern hemispheres. This would, therefore, be the mechanism that could have generated the cycles of 41,000 years in the period preceding 1 million years ago.

The variations in eccentricity can, in turn, lead to variations in summer insolation. The greater the eccentricity, the greater the difference between summer and winter insolation. In addition to this, the differences in insolation between the two hemispheres depend on the direction of the axis of rotation. If the axis of rotation 'leans' towards the inside of the orbit when the Earth is at its closest point to the Sun (perihelion), then in the summer, the Northern hemisphere will receive much more energy than the Southern hemisphere (and vice versa). Together, these fluctuations in orbital parameters can cause a variation with a period of around 100,000 years in the summer insolation of each hemisphere. In any case, the variations in the eccentricity of the Earth's orbit remain so small that the effect of any variation in obliquity dominates.

Despite this, a purely astronomical explanation is not enough. The variations in energy received are modest, and they are not sufficient on their own to explain the great climatic differences that exist between glacial maximums and interglacial periods, or the sudden passages between one state and another that we have seen during the most recent deglaciations. Furthermore, we must understand why the glacial oscillations activated only 3 million years ago and why, around a million years ago, the glaciations went from a period of 41,000 years to a period of about 100,000 years.

Glacial Mysteries

In popular articles and books, we often hear the phrase 'revealing the mysteries of the ice ages'. This is actually an extremely complicated issue on which we are starting to have huge amounts of data and more articulated ideas that challenge us to substantiate our understanding of the climate system and the relationships of cause and effect – and the effects that, in turn, become causes.

First, we have to consider the trigger for these oscillations. Around 5 million years ago, the global temperature was roughly 3 °C higher than it had been in subsequent millennia, and the carbon dioxide content in the atmosphere was around 400 ppm, higher than the typical levels found before the industrial era (though comparable with today's). There was ice in the Antarctic and a reduced ice sheet in Greenland, but we have no evidence that there was any glacial extension in Eurasia or North

America. The water cycle was more intense, and at the tropics and the equator there were probably permanent *El Niño* conditions, meaning no alternation between the two states of *El Niño* and *La Niña* that we see today and which we discussed in chapter 9.

The climate, however, was in the throes of the descending trend, in terms of both CO_2 and temperature, that had begun millions of years earlier. At a certain point, the continual decrease in carbon dioxide that characterized the Cenozoic caused the temperature to drop enough to trigger the formation of large extensions of ice on the continents located at high northern latitudes. The thermohaline circulation was reorganized and the dynamics of the equatorial Pacific began to oscillate between the two alternative states of *El Niño* and *La Niña*.

Furthermore, at that time, the movement of continents led to small yet significant changes in the layout of the emerged lands and, consequently, in marine circulation. For example, the Isthmus of Panama was formed, which by joining North and South America closed the equatorial opening between the Atlantic and Pacific oceans. All of this, together with changes in the water cycle that led to greater precipitation and the progressive increase in ice on the continents of the Northern hemisphere, made the Earth System more sensitive to changes in obliquity, bringing it to a point of instability, wherein a small external perturbation could lead to the appearance of an oscillatory behaviour.

And so the cycle of 41,000 years was set in motion, with the weak astronomical force strengthened by a series of mechanisms internal to the climate's dynamics. Firstly, the albedo of the ice and snow amplified the variation in temperature, modulating the quantity of solar radiation reflected and absorbed. The weight of the forming ice then led to adjustments in the continental crust that could have diminished volcanic activity, reducing greenhouse gas emissions in the accumulation phase and increasing them when the ice melted.

Similarly, the marine ice reacted rapidly to variations in temperature, reinforcing them and modifying the exchanges of energy and water between ocean and atmosphere. But there was also a response from the marine ecosystems that influenced the amount of carbon dioxide in the atmosphere by increasing the productivity of phytoplankton and the absorption of CO_2 through photosynthesis, followed by the burial of organic matter in the deepest part of the oceans. Here, weaker ocean mixing in the glaciation phase led to an accumulation of organic matter, slowing its return to the surface, further reducing the atmospheric concentration of carbon dioxide and, eventually, the greenhouse effect. In short, the whole system had entered into an oscillatory state in step with the astronomical force and the breath of the ice with its expansion and reduction.

The World of the 100,000 Years

Around 1 million years ago, something else happened that triggered a climate response with a period of around 100,000 years. It would seem that the variations in eccentricity were not as pronounced as those in obliquity, and the possible explanations call into question the conjoined action of at least two different mechanisms. The first, proposed once again by Raymo and her team, is that the decrease in atmospheric CO_2, which also continued during the Pleistocene, could have led to ever lower temperatures and a progressive increase in the global quantity of ice between one cycle and another. The oscillations of 41,000 years did not lead to interglacial periods with the same conditions each time, but to an ice cover that, though reduced in comparison to the Glacial Maximum, was nevertheless a little more extensive than during the previous inter-glacial period. In the long term, the rise in summer insolation due to changes in obliquity was not sufficient to melt the ice, and the period of the cycles was lengthened as a result.

Alternatively, paleoclimatologists Peter Clark and David Pollard identified in 1998 a possible cause for the transition in geomorphological changes induced by the ice sheets themselves and linked to the thickness of the sediments beneath the ice. If there is a thick layer of sediment beneath a glacier, flow is facilitated towards the area where it will then melt. However, each time a glacier advances, it partially erodes the sediment beneath it and, over the course of millennia, all the sediment can be removed as a result of this erosion. At that point, the glaciers are left sitting on compact rock, which slows their movement and their eventual melting. In this way, the accumulation of a greater thickness on a given glaciated surface becomes possible, allowing for a net increase in the global volume of ice with the ensuing effects on the thermohaline circulation. It might be for this reason that the period in which ice was present grew longer, bringing a new situation characterized by a periodicity of 100,000 years.

It is also possible that the cycles over the last million years, which do not actually have a fixed duration, are not an oscillation of 100,000 years but rather a mixture of oscillations between 80,000 and 120,000 years, meaning a glacial cycle is repeated every two or three periods of obliquity. This interpretation, proposed by Peter Huybers in 2009, is based on the idea that the previous mechanisms may have effectively caused a glacial melting to be 'missed' every so often and that the climate system's response became more chaotic, without the same periodicity as the force. Starting from a million years ago, in the phase of maximum insolation and maximum obliquity, some deglaciations happen and others do not. One possible explanation for this behaviour is linked to the lack of stability inherent in the ice sheets: the bigger they are, the more they tend to collapse. In the 1980s, a similar mechanism was theoretically explored by paleoclimatologists Michael Ghil and Hervé Le Treut, from UCLA and the École normale supérieure in Paris, combining the thermodynamic balance of energy with the dynamics of the cryosphere and the Earth's crust in an approach that would later be called 'cryothermodynamics'. Many theoretical models have since been proposed to describe the almost regular alternation between Glacial Maxima and interglacial periods in the last 3 million years, but we have not yet attained a full understanding.

As normally happens in many non-linear systems where the response is not proportionate to the stimulus, for the Earth's climate the combination of a slow decrease in atmospheric CO_2 due to geological reasons and an approximately periodic astronomical perturbation could have caused the climate to go through a series of successive transitions. Starting with the dynamics of non-linear systems, in 1982 Italian theoretical physicists Roberto Benzi, Giorgio Parisi (awarded the Nobel Prize for Physics in 2021), Alfonso Sutera and Angelo Vulpiani, and Belgian climatologist Catherine Nicolis independently, put forward a conceptual mechanism with which to explain how the climate system could have generally amplified an extremely weak external perturbation such as the variation of astronomical parameters. The basic point is that, when a non-linear system with two possible states (glacial and interglacial) is simultaneously compelled by a weak periodic external force and a random (stochastic) internal 'noise' – generated, for example, by atmospheric and ocean motions – it is possible to create a resonance between the characteristics of the noise and those of the periodic force that leads the entire system to amplify the external perturbation massively and enter into a condition of continuous oscillation between the two states. This mechanism, known as stochastic resonance – whose validity for effectively explaining the glacial–interglacial transitions has yet to be verified – has nevertheless given rise to studies and applications in the most disparate sectors of science and technology, including the construction of extremely efficient amplifiers.

We have learnt a lot about the mysteries of the ice age, but there is still a great deal left to understand, particularly when it comes to the sequences of cause, effect and response by various components of the climate system. We have also found confirmation that carbon dioxide, together with the albedo, almost always plays a crucial role, but that often its responses are modulated and complicated by many other actors, including the global volume of ice. Over the next few years, the EPICA project will continue its work, leading to deeper coring and reaching even older times, giving rise to new analyses that will allow us to understand better the fluctuations between Glacial Maxima and interglacial periods, and the global response of the climate.

I would like to end this chapter with an observation that I will return to shortly – namely, that this is the world of glacial fluctuations in which we have evolved, and that has always been characterized by atmospheric concentrations of carbon dioxide below 400 ppm. We have to go back millions of years to find higher levels. Either that, or we can look at the concentrations over the last few decades, which have returned to what they were more than 3 million years ago. We will shortly look at what this might mean, though we, or our descendants, might well witness it personally in the near future.

12

Agitated Ice

In the middle of the last ice age, especially between 60,000 and 20,000 years ago, the climate was particularly variable, showing significant instability that put the survival of human beings to the most extreme of tests. The old, arid world of hunter-gatherers, who were distributed throughout the continents developing language, music, art and the first metaphysical or religious ideas, was on the verge of ending. The last ice age was moving towards its conclusion, and the ice would soon retreat, leaving in its wake a different planet with a milder climate. A planet ready to be explored, populated, invented.

The Glacial Maximum

Around 20,000 years ago, ice covered most of Northern Europe and North America. This was the maximum of the Würm glaciation, as it is known by geologists – the one that left the clearest and best-preserved marks on our current landscape. Canada and the Northern United States were covered by an ice sheet called Laurentide, which came as far south as Boston, with the sandy peninsula of Cape Cod today marking its edge on the Atlantic Ocean. Scotland was almost entirely covered by hundreds of metres of ice and only the highest Alpine peaks could be seen emerging from the glaciers.

The sea level was almost 100 metres lower than it is today, because an enormous amount of water was trapped in polar and mountain ice glaciers. As a result, vast coastal regions were revealed and the South of England was connected to the European continent. The Bering Strait became a broad isthmus linking Siberia to Alaska, partially covered by ice, but also, most likely, had coastal areas that were open and useable. It is believed that it is through here, towards the end of the glaciation, that Asian populations reached North America and then, in a complex series of migrations, made their way as far down as Patagonia, using the incredible capacity to adapt to almost any environmental conditions that characterizes our species. The hunters from Clovis culture, named in reference to the town in New Mexico (USA) where their remains were first found in the 1920s and 1930s, populated the Americas more than 13,000 years ago, though remains of even older humans (known as 'pre-Clovis') have been discovered in the last 20 years, and we are not yet sure of the exact paths taken by these first inhabitants to reach the American continent. In any case, it is mind-boggling to think of the Clovis people migrating from Siberia to America through the Bering Strait, probably over several generations and driven by the search for new lands and more game, through unexplored territories with the foaming ocean on one side and vast walls of ice on the other.

An Unstable Climate

Marine and polar cores tell us the glacial climate was rather unstable. It was generally much drier and undoubtedly colder than today, even if in some regions that are currently semi-desert (like the west of the United States), the change in atmospheric circulation caused greater rainfall. The climate was, however, subject to fairly rapid variations over the course of thousands of years.

In 1988, the young German marine geologist Hartmut Heinrich published a surprising result: the oceanic sediments at the bottom of the Atlantic revealed various layers of grit and sediment of continental origin, known as ice-rafted debris and transported by glaciers. The interpretation provided by the researcher was that they had been generated by spectacular iceberg releases due to collapses in the Arctic ice sheet. The icebergs carried large quantities of trapped sediment, which sank to the seabed when the ice melted once it reached the warmer waters at lower latitudes. And so, 'Heinrich events' were discovered. Six were identified during the last glaciation, happening at intervals varying between 5,000 and 6,000 years.

To date, two possible causes for this oscillating behaviour have been put forward. The first is that the Greenland ice sheet, growing steadily due to the low temperatures at that time, would periodically reach a state of instability, suddenly releasing portions of its more external ice before slowly reforming and starting the cycle once more.

Alternatively, it is possible that the thermohaline circulation (which we have already discussed in chapter 7) was subject to a series of oscillations by the release of fresh water from the melting land ice, thus causing it to fluctuate between two states: the one we see today, and an alternative state of weaker circulation generated by the fact that the less dense fresh water released into the Arctic by melting ice sharply reduced the sinking of surface water, decreasing the formation of deep water. We will return to this shortly.

In order to understand better what happened, we must consider that there were other significant fluctuations in the glacial climate. Since the 1980s, the analysis of the cores of marine sediments and ice in Greenland has revealed the presence of sudden climatic changes during the ice age. Analysing the ice cores extracted in Greenland, called GRIP and GISP2 (acronyms of the corresponding research projects), Danish paleoclimatologist Willi Dansgaard, Swiss climatologist Hans Oeschger and their team were able to pinpoint precisely a large number of rapid hot–cold variations in the isotopic signal of oxygen-18.

These fluctuations, now known as Dansgaard–Oeschger oscillations, begin with a sudden warming that might last only a few decades and seemingly leads to interglacial conditions, followed by a very slow cooling, or by warm conditions that last for several centuries, concluded by a rapid cooling of the water that causes the return to glacial conditions. In the last glaciation, starting from around 120,000 years ago, 25 such events have so far been discovered, some more intense, some weaker and shorter. The strongest fluctuations also correspond to Heinrich events, which belong to the same context of climate variability. Figure 12.1 shows the temperature signal estimated from the isotopic ratio of oxygen-18 to oxygen-16 obtained from the GISP2 glacial core, for the last 50,000 years.

Global Fluctuations

Given the rapid fluctuations in temperature registered in Greenland, we might ask ourselves what was happening in Antarctica during the same period. Here too, the ice cores reveal the presence of climate fluctuations on temporal scales of thousands of years, similar to the Dansgaard-Oeschger events in the Northern hemisphere, albeit less sudden and

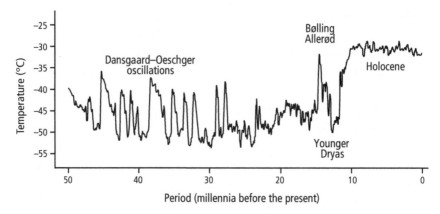

Figure 12.1 The temperature record, estimated from the isotopic ratio of oxygen-18 to oxygen-16 obtained from the GISP2 perforation in the Greenland ice. We can clearly see the rapid Dansgaard–Oeschger oscillations and the warming from 14,000 years ago, interrupted by the event known as Younger Dryas (around 13,000 years ago) and followed by the transition to the warmer conditions of the Holocene a little less than 12,000 years ago. This record was taken from the work of Richard Alley (2000) and refers to the temperature in central Greenland.

intense. In particular, the Antarctic record reveals a slow increase in temperature followed by a more rapid return to cold conditions, the opposite of what was seen in the Arctic. In order to understand whether the two kinds of fluctuations are connected, and whether the two hemispheres vary simultaneously or in opposite phases, it is necessary to obtain a precise synchronization of the ice cores obtained at the two poles.

The conjoined dating of the various ice cores happened at the beginning of this century, thanks to the variations in methane concentration measured from the ice-trapped air bubbles. Methane variations are a global phenomenon, equally present in both hemispheres, and thus allow for a synchronization of the Arctic and Antarctic ice cores. In 1998, a group of climatologists led by Thomas Blunier, who at the time was working in the US, used the methane method to demonstrate that the Arctic and Antarctic oscillations were asynchronous. Blunier and climatologist Edward Brook later demonstrated that, for the most intense events of the last 90,000 years, the phases of a slow drop in

Arctic temperatures correspond to a slow rise in Antarctic temperatures, while the sudden Antarctic cooling events correspond to the rapid warming events measured in Greenland. So, the two hemispheres have opposing phases when it comes to rapid glacial fluctuations. In 2006, the working group of the EPICA paleoclimatic research project extended the analysis to all Dansgaard–Oeschger events, even the weakest, demonstrating that in these cases also there is an anti-phase correspondence between temperature fluctuations in the Arctic and the Antarctic. The Southern hemisphere, however, responds with a delay of several decades, a behaviour that indicates a slower reaction by the southern oceans.

The Bipolar Seesaw

The origin of the bipolar oscillations during the glacial period continues today to be the subject of discussion, but the mechanisms proposed are generally linked to the variability of the thermohaline circulation, which we looked at earlier. In general, the very cold and salty water sinks in the Arctic and then moves to the depths of the Atlantic Ocean, first towards the lower northern latitudes, before then moving past the equator and flowing towards the mid and high southern latitudes. This recalls warmer superficial waters that come partly from the Southern hemisphere and move heat towards the Arctic. Already in 1870, Scottish climatologist James Croll (whom we met when talking about Milankovich cycles) noted that the temperature of these southern surface waters was higher than that of the deep waters coming from the north, leading to a net heat transfer through the equator from the Southern to the Northern hemisphere.

If the thermohaline circulation were to weaken, the effect would be a reduction in heat transport through the equator, encouraging it to accumulate at the Southern hemisphere. This would eventually lead to a cooling of waters at high northern latitudes and a warming of the southern waters, as already proven by Scottish paleoclimatologist Thomas Crowley and the group coordinated by Swiss climatologist Thomas Stocker in 1992, and revisited by Wallace Broecker, whom we have already met, in 1998. Specifically, it was Broecker who coined the term 'bipolar seesaw' to describe this mechanism.

The mechanism functions more or less like this, as described by Stocker and Icelandic-Danish glaciologist Sigfús Johnsen in 2003: if, at a certain moment, due to a warming at high latitudes or to the instability of the Greenland ice sheet, there is an intense melting of the ice and eventually the detachment of several icebergs (which, in the most extreme cases, would generate a Heinrich event), then the fresh water released on the surface renders the water less dense, reducing or stopping it from sinking. As a consequence, the thermohaline circulation slows and is weakened – in the most extreme cases, it might even collapse – and the Arctic water no longer moves through the oceanic depths towards tropical and then southern latitudes. In parallel, this reduces the return of warm surface water from the south, meaning the temperature of the Arctic water tends to decrease.

In the Southern hemisphere, the reduction of the northward heat transport allows heat to accumulate and consequently causes a rise in temperature at all southern latitudes and in Antarctica. Here, however, the process is slower due also to the lesser extension of the continents and the structure of the southern oceanic currents.

When the Arctic is finally sufficiently cooled and the melting of continental ice has reduced, the now saltier marine waters begin to sink once more and the thermohaline circulation picks up again, returning to its 'normal' state. The transport of heat from the Southern hemisphere restarts and the temperature in the southern oceans and in Antarctica drops once more. The Arctic seas, on the other hand, begin to warm again until they eventually trigger a new cycle of ice melting and another collapse of the circulation. This is, therefore, a possible origin of the bipolar oscillation that connects – in opposite phases and with a delay caused by the characteristics of the southern oceans – the variations in temperature measured in the Arctic and in Antarctica, connecting them to a global instability of the thermohaline circulation. It has also been suggested (though the observational evidence is modest and the conclusion hotly debated) that this oscillation repeats itself cyclically with an approximate period of around 1,470 years.

Naturally, the explanation that has just been given is only a conceptual model, as many aspects are yet to be understood in depth. A central role is played by deep ocean circulation, but there are crucial interactions with the dynamics of ice sheets and there can be significant influences from the atmosphere, fluctuations in carbon dioxide and the response of ecosystems. The variations in Arctic marine ice cover in response to temperature fluctuations also play an important role. For anyone wanting to know more about these aspects, the wonderful article published by climatologist Laurie Menviel (who currently works in Australia) and her co-authors provides a wide-ranging overview of this fascinating issue.

The Dansgaard–Oeschger oscillations and Heinrich events are not an anomaly of the last glacial cycle. Similar cycles have been identified, albeit with greater difficulty, in previous glacial periods over the last 800,000 years. This would appear, therefore, to be a typical way in which the climate functions in glacial conditions. Even more interesting is the observation that such variability in the thermohaline circulation could continue in a more nuanced way through interglacial periods and could, therefore, be present today. Over recent years in particular, the possibility of a slowing down of the thermohaline circulation and the Gulf Stream due to intense glacial melting and Arctic warming has been widely debated. This change in deep Atlantic circulation is often cited as one of the climate's so-called tipping points. Or it could also be a manifestation of the bipolar seesaw, its natural dynamics being disturbed by global warming of anthropic origin.

The Ice Retreats

Around 15,000 years ago, mountain glaciers and ice sheets were in rapid retreat. The sudden end of the last ice age had arrived, in a similar way to how it happened many other times over the last 3 million years. Enormous quantities of ice melted and caused the sea level to rise rapidly by around 80 metres in a just few thousand years. Huge quantities of sediment were deposited, transported by the meltwaters, and the morphology of the coasts changed drastically. Almost all the components of the climate were modified, with the movement of plants and animals and significant changes in ecosystems all over the planet. Gigantic lakes formed on the edges of retreating ice sheets. Lake Agassiz in North America, for example, at the foot of the Laurentide ice sheet in Manitoba, Canada, was probably as large as the Black Sea.

The period ranging between 20,000 and 12,000 years ago is a mine of information for understanding the climate's response to external forces and the mechanisms of its internal instability, and helps us understand what could happen when the temperature quickly rises. During those 8,000 years, due to variations in orbital parameters, the insolation at 65° north in the month of June rose from 420 to more than 460 W/m². In a wide-reaching 2012 review article based on statistical data analysis, a group of researchers, coordinated by US paleoclimatologist and glaciologist Peter Clark, identified two key mechanisms by which the climate left the Last Glacial Maximum.

The first, on a global scale, concerns the temperature's response to a rise in concentration of greenhouse gases. The second, responsible for differences in response times in the two hemispheres, involves the oscillations of the thermohaline circulation, that same process responsible for rapid oscillations during the glacial period. As well as this, the melting ice liberated new areas, which probably became boggy, absorbed more sunlight, grew warmer and, in turn, emitted carbon dioxide and methane.

The relationship between the rise in temperature and carbon dioxide is, of course, an important one. At the beginning of the interglacial period in which we are currently living, the concentration of carbon dioxide in the atmosphere rose from little more than 180 ppm to around 260 ppm, in step with the rise in temperatures. As discussed in the work of US paleoclimatologists Jeremy Shakun, Peter Clark and various others, the data suggests that the increase in the concentration of carbon dioxide happened before the rise in global temperatures, which rose with a delay of several hundred years. But there are other significant aspects to be considered.

In Antarctica, for example, the temperature rose almost simultaneously with CO_2 levels, meaning it preceded the variation in global temperatures. Particularly at the beginning of deglaciation, the temperature rose just over half a degree before the increase of carbon dioxide in the atmosphere. The picture that emerges is therefore a slight rise in temperatures, caused perhaps by a rise in summer insolation, which unleashed a collection of powerful climate responses, including a rapid rise in the concentration of atmospheric carbon dioxide, released by the oceans or the increased productivity of ecosystems. This, in turn, led to a further rise in global temperatures and an acceleration of deglaciation.

It is also possible that the collapse of the Arctic ice shelf, due to the same mechanisms we have already discussed with the Dansgaard–Oeschger oscillations, had initially slowed the thermohaline circulation, enabling a rise in Antarctic temperatures as a first effect. However, this time, the climate system did not continue with its usual oscillations, as had happened during the previous period, but created something new. There was a spasmodic fluctuation and this signalled the end of the last glaciation.

Dryas!

The mountain avens (*Dryas octopetala*) is a small plant that lives in the cold environments of the Arctic and high mountains, and the presence of its pollen, for example in ancient sediments at low elevation and more southern latitudes, is indicative of glacial conditions. Between the end of the nineteenth and the beginning of the twentieth centuries, Scandinavian botanists discovered Dryas pollen in peatbogs, at the bottom of lakes and in sediments in Denmark and Sweden in strata formed around 12,000 years ago. This was an indicator of very cold conditions that had interrupted the general move towards deglaciation and which were later confirmed by numerous cores and climate reconstructions.

Between less than 13,000 and 11,700 years ago, in full deglaciation, the climate suddenly returned to glacial conditions, leading to the expansion of polar species at lower latitudes. This period was called 'Younger Dryas' in order to distinguish it from two other cold periods, similarly associated with the expansion of the mountain avens, which happened a few millennia earlier and are known respectively as 'Older Dryas' and 'Oldest Dryas'. In any case, the Younger Dryas was the most obvious example, and with the most significant consequences, of a return to glacial conditions. Even in the Alps at that time, the icy blankets advanced once more before retreating over the following millennia.

Understanding what generated this episode is, obviously, important because of the implications it can have for the current climate. There is no lack of hypotheses, such as the possibility of climate change caused by the impact of a huge meteorite, intense volcanic eruptions, or even the explosion of a supernova. All in all, none of these somewhat ad hoc explanations has stood up against the data.

Rather, we have to return to the variations of the thermohaline circulation, interpreting the Younger Dryas episode as the last great instability similar to that of the Dansgaard–Oeschger oscillations. In this case, one of the causes of instability could have been a change in drainage conditions in Lake Agassiz, which around 13,000 years ago led to the release of a large quantity of fresh water into the oceans at high latitudes, presumably altering the sinking of the Arctic waters and triggering a chain of processes that we have learned to recognize.

Slowly, the thermohaline circulation restarted its work of transporting heat towards the north, and, around 11,700 years ago, the climate left the glacial conditions of the Younger Dryas. At that point, the glaciation had truly finished and the Holocene began – the era of humanity, agriculture, of the birth of cities, kingdoms and empires, philosophy, art and science.

13

Conquering the Planet

The end of the last glacial peak saw the arrival of a warmer and rainier period characterized by a relatively stable climate compared to the significant oscillations of the previous millennia. Humankind had already spread throughout the planet thanks to migrations that began tens of thousands of years earlier, and had started developing agriculture and farming practices. This made it possible to produce greater quantities of food, a situation that allowed some individuals to dedicate their time to activities that went beyond their immediate survival. Society became more structured and economic resources became concentrated in ever fewer hands. New ways of life emerged, such as those of philosophers, soldiers, officials, professional kings and emperors, marking the beginning of the age of human history, with all of its splendours and tragedies.

The Climate of the Holocene

The Holocene, the current geological epoch (or the penultimate according to some, a view we will discuss in the next chapter), is conventionally held to have begun about 11,700 years ago with the planet's emergence from the glacial conditions of the Younger Dryas. From then until now, the climate has continued to vary, albeit in a much less violent way than it did during the ice age.

Over the course of the years, many research groups have estimated the variability of the temperatures and precipitations of the last 12,000 years using various sources of information: the chemical characteristics and isotopic ratios obtained from marine and lacustrine sediments, peat bogs and ice cores; the distribution of biological species and pollens; microfossils; the analysis of speleothems; the characteristics of tree rings; and much more.

In 2020, the vast research group coordinated by US geologist and climatologist Darrell Kaufman produced a summary of the data from

679 different sites as part of a colossal work that provided a comprehensive overview of temperature variations in the Holocene. Figure 13.1 shows the variations in global temperature over the last 12,000 years as ascertained by this work, alongside another two reconstructions published a few years earlier. We can see how, since the end of the ice age, global temperatures have varied by 1 °C at most, even if there were more intense geographically localized fluctuations. We have therefore had a much more stable climate that has allowed for the development of agriculture and settled human civilizations all over the planet.

The Green Sahara

In the period between around 9,000 and 5,000 years ago, global temperatures were higher than in the millennia that followed immediately. This event, called the Holocene Climatic Optimum, was predominantly localized in the Northern hemisphere and particularly evident at higher

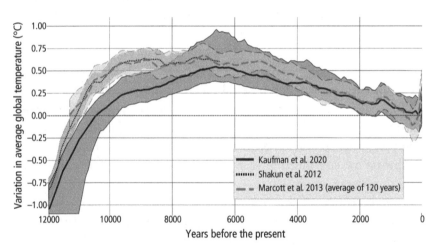

Figure 13.1 The continuous line shows the reconstruction of average global temperatures in the Holocene as ascertained by the work of Darrell Kaufman et al. (2020), and is compared with another two recent reconstructions. The grey areas provide an estimate of temperature uncertainty. The temperatures are presented as deviations (in degrees centigrade) from the average temperature in the period 1800–1900 CE.

latitudes. In Greenland, for example, the temperature was around 3 degrees higher than in the last 4,000 years. Around the Baltic Sea, fossil remains indicate the change from tundra, which was present during the period directly after the glaciation, to oak forests, typical of higher temperatures, and then again to beech forests, indicating a return to a cooler climate.

The most likely cause of this heating is the variation in terrestrial orbital parameters, in particular the obliquity, which led to greater summer insolation at high northern latitudes. The astronomical effect was then amplified by the response in the Arctic regions, with a decrease in albedo caused by a reduction in continental and marine ice.

In that period, the Sahara Desert was rainier than it is now, leading it to be referred to as the 'green Sahara'. In 2017, paleoclimatologist Jessica Tierney at the University of Arizona and her colleagues published a detailed reconstruction of precipitation in North-Eastern Africa based on the analysis of the isotopes present in the external part of the leaves contained in marine sediments near the coast. The study's results suggest the probable presence of permanent lakes, human settlements and a rich and diverse vegetation in the Sahara area.

In the case of the Sahara, the higher levels of rainfall were presumably generated by changes in the surface temperatures of the sea, as well as a different distribution of dusts in the atmosphere and the positive interaction between vegetation and precipitation (already discussed in chapter 6), with corresponding changes in the monsoon circulation. A rise in precipitation led to much more abundant vegetation, which in turn increased rainfall on a regional level. It was essentially a large-scale Charney mechanism, involving albedo, evapotranspiration, wind regimes and atmospheric dusts.

This warm period broadly coincided with the Neolithic period and the first structured human settlements, which culminated in the great Sumerian cities such as Uruk around 6,000 years ago. The bonds between climate and the development of human civilizations are fascinating, but we mustn't fall into a mechanistic way of seeing things and explain the birth, death and development of civilizations using only external causes such as changes in climate, when the dynamics within human societies play such a central role. Nevertheless, it is no doubt intriguing to think of how the end of the climatic optimum and the aridification of the

Sahara could have contributed to the myth of human banishment from the Garden of Eden.

The Last 2,000 Years

Moving into historical times, we have more and more information on climate variability, including direct accounts – which are not always easy to interpret because they are often influenced by the social and cultural context – and paintings that depict the environments of the periods in which they were created. But we can also glean information from tree rings, high temporal resolution marine cores, peat bogs and so on. The PAGES 2k (Past Global Changes) international research consortium aims to collect the results of these reconstructions in order to provide a complete picture of everything we know about the climate over the last 2,000 years.

What emerges is a relatively modest variability with an alternation between warmer and cooler periods, but with deviations in average global temperatures that rarely go beyond half a degree Celsius (warmer or cooler) than the typical temperatures seen in the period between 1850 and 1950, albeit with notable differences between the different parts of the planet. Figure 13.2 shows the reconstruction of global temperatures over the last 2,000 years obtained from PAGES 2k. Here, we can see a 'slow' variability that is not insignificant, with medieval temperatures just a little lower than current ones (until the early centuries of the second millennium CE), followed by a slow decrease (the so-called Little Ice Age) interrupted by very rapid heating in the twentieth century. Naturally, these are reconstructions like all the others that we have seen in the previous chapters. There were no mercury thermometers in Roman times. New data, new cores and better data analysis can still change the details of this picture, particularly when it comes to single continents, but the overall trend is clear.

Information on the climate of the past comes, as we have said, from many different sources. As well as isotopes, we can also use information provided by the characteristics of tree rings, with the sciences known as dendrochronology (used for dating, often obtained using the carbon-14 method) and dendroclimatology (used for actual climate reconstruction). The latter discipline makes the most of the fact that the thickness of the

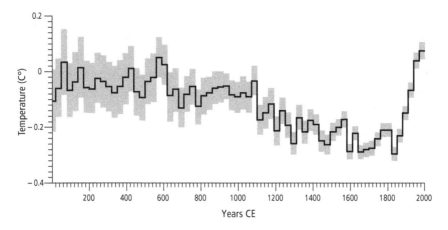

Figure 13.2 Reconstruction of global temperature fluctuations over the last two millennia, represented as the difference with respect to the average of the period ranging from 1850 to 1950, created by the international research group PAGES 2k using measurements from various climate archives the world over. The dark curve indicates the average and the shaded area indicates the uncertainty calculated by the entirety of available measurements. Indicated here are 30-year averages. This work has allowed for an accurate reconstruction of temperature trends important for comparisons with the dynamics of the current climate. Redesigned by www.nature.com/articles/sdata201788, the data is available on the NOAA website, www.ncdc.noaa.gov/paleo-search/study/21171.

trees' annual growth rings and the characteristics of the wood register the climatic conditions (temperature, precipitation, drought conditions, fires) at the time the ring was formed. This information combines various factors and it is not always easy to distinguish one cause from another. For example, we often see a sign that mixes temperature and precipitation. Furthermore, each species of tree can respond differently and it is therefore necessary to know the subjects in question extremely well.

At the moment, the oldest trees dated with precision lived more than 12,000 years ago. One particularly famous tree in this context is the Bristlecone Pine (*Balfourianae Engelmann*), extremely long-living and very slow-growing, which can be found in the arid, high-altitude zones of the western United States. One of the oldest (still living) is called Methuselah and is more than 4,800 years old, as indicated on the online

list of the oldest trees that can be found, on the website www.rmtrr.org/
oldlist.htm. Often, however, the chronologies span shorter periods due
to the difficulty in finding samples of old wood suited to dendrochronological study. Particularly important here is the analysis of trunks used
in the construction of ancient buildings, which are often well conserved
and suitable for use in climatic reconstructions.

As well as studying global conditions, many research groups have
analysed the climate history of specific regions of our planet. Figure 13.3,
for example, shows an estimate of summer temperatures (from June
to August) for Scandinavia and the European Alps, obtained by paleoclimatologists Ulg Büntgen and Willy Tegel, who combined five different
dendroclimatic reconstructions, and published as part of the PAGES
consortium's activity that we have already referenced. The figures also show
the climatic anomaly of the Middle Ages, to which we will return shortly.

In 2020, US hydroclimatologist Park Williams, together with a wide
range of co-authors, published an estimate of soil humidity in the
southwest US for the last 1,200 years, combining information taken from

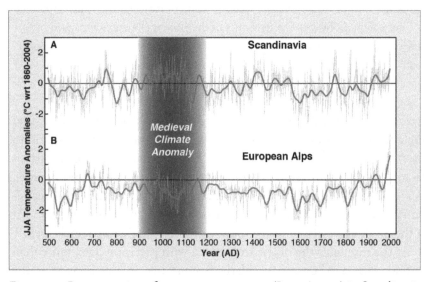

Figure 13.3 Reconstruction of summer temperatures (June–August) in Scandinavia
and the European Alps, combining dendrochronological and dendroclimatic
information from five different reconstructions. Taken from Büntgen and
Tegel (2011).

a significant number of dendroclimatic reconstructions with the results of hydrological simulations, as shown in figure 13.4. This work revealed how the years dating from 2000 to 2018 constituted the driest 19-year period since the end of the sixteenth century, and it showed how, over recent decades, there has been a growing tendency towards situations of 'mega-drought'. Climate change is, therefore, beginning to show its teeth, as we will discuss over the coming chapters.

Finally, the closer we get to the present age, the more information is provided by summaries, reports and paintings, and the more evident the interconnection between the variability of the climate and human history. A great classic in this field is the book *L'Histoire du climat depuis l'An Mil* by French historian Emmanuel Le Roy Ladurie, which tells the history of the climate from the year 1000 and how this has influenced the development of human societies. PAGES 2k have collected an enormous number of documents on historical climatology (as this discipline is known) in a special volume dedicated to historical archives. It is also well worth remembering that climatologist Dario Camuffo of the Italian National Research Council was one of the pioneers in the study

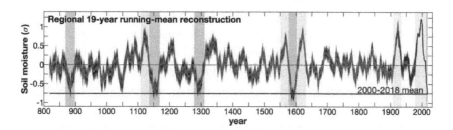

Figure 13.4 Estimate of soil humidity in the south-western United States for the period from 800 to 2018 CE. The data before 1900 CE come from dendroclimatic and modelling reconstruction, while more recent data include also direct measurements. The thin line in the centre indicates the average for the entire period and the thicker horizontal line at the bottom is the average for the period 2000–18. The periods highlighted in lighter grey indicate those that were particularly humid, whilst those shown in dark grey are the periods that were particularly dry. Taken from https:// news.climate.columbia.edu/2020/04/16/climate-driven-megadrought-emerging -western-u-s and redesigned by Park Williams et al. (2020).

of historical climatology and the interpretation of the scenes depicted in ancient paintings in order to glean information on historical environmental conditions.

Romanesque Heat and Baroque Cold

Over the last 2,000 years and before our own century, there were two particularly relevant periods from a climate perspective: the Medieval Warm Period, which in reality was also present in the preceding centuries and particularly evident around the year 1000 and in the centuries that immediately followed, and the Little Ice Age between 1500 and 1800.

The Medieval Warm Period has been the subject of much discussion, often more similar to diatribes from the supporters of rival soccer clubs, rather than a careful analysis of objective data. What we know – for example, from the results of PAGES 2k – is that, until around the twelfth or thirteenth century, the world was warmer than in later centuries, but with notable geographical differences and, in some cases, notable temporal variations. This warm period was particularly evident in North America, where it may have led to the decline of the Anasazi society in the south-western United States.

The data for this period in South America does not show a particularly warm climate, while in Europe the temperatures were higher than in the centuries that immediately followed but also much more variable, with significantly more marked temporal fluctuations over just a few decades, to the extent that researchers have recently started referring to a 'Medieval Climate Anomaly', rather than to a completely warm period. Some historians have linked the exploration of America and the colonization of Greenland by the Vikings to more favourable climate conditions, along with the cultivation of grape vines in Britain and the presence of olive trees farther north than they would usually grow. But there are various other social and economic causes, including wars, that could have influenced these events and choices (for example, it would have been easier to put up with drinking a mediocre wine than to import a better one from southern regions). Overall, the Medieval Warm Period seems to have been a characteristic of the area around the North Atlantic (Europe and North America) rather than a global phenomenon. In figure 13.1, we can also see how this period relates (in the form of a positive temporary fluctuation) to the general trend towards decreasing temperatures that followed the Holocene Climatic Optimum. But it is beyond doubt that those centuries were, on average, hotter than what was to follow.

From the thirteenth century, global temperatures began to drop, reaching particularly low levels in Europe around 1600. The time of the Little Ice Age had arrived, characterized in Europe by cold and humid winters and the extensive freezing of rivers and lakes, celebrated in numerous paintings such as those by Pieter Bruegel the Elder, or those held in the Museo Correr showing the Venice lagoon frozen over.

The extremely harsh and long winters created enormous problems for the world's populations, and the sudden freezes often destroyed harvests, leading hundreds of farming families to die from starvation. Superstition began to reign supreme and a guilty party was sought for nature's 'mutiny', which was otherwise incomprehensible and so perceived as an unjust punishment. And the guilty party was soon found in the supposedly demonic work of witches and warlocks. Thousands of trials for witchcraft saw some men, though mostly women, accused of being responsible for hailstorms and cold snaps. There was a wide-scale witch hunt during this dark and fearful time, in which harvests were destroyed and what was salvaged then rotted in the stores. *The Devils*, a beautiful and visionary film by Ken Russell, tells only too well the story of that desperate world. The classic book *Climate, History and the Modern World* by English climatologist Hubert Lamb, the book *The Little Ice Age: How Climate Made History* by British anthropologist Brian Fagan, and the most recent book by German historian and journalist Philipp Blom, *Nature's Mutiny*, all explain how European societies reacted to the great cold and which developments came about as a result of those difficult times.

During the Little Ice Age, glaciers on the Alps stretched down towards the valleys. This was nothing in comparison to the enormous expanses of ice some 20,000 years before, but enough to leave moraines and signs of expansion that are clearly visible even today. Elsewhere, traces of the cold period are less evident, raising the question of whether it was truly a global phenomenon or whether it was essentially regional like the Medieval Anomaly, predominantly present in the areas around the Atlantic – Europe in particular. At the moment, clues seem to indicate a localized nature for this cold period, though, as ever, only new research and new quantitative data can answer such questions.

Does the Sun Have the Answer?

The obvious question, now that we have explored the great climate changes of the past, is what caused the medieval warmth and the Little Ice Age. The first interpretations of these fluctuations in climate focused their attention on solar activity: not variations in terrestrial orbit, as in the case of glaciations, but changes in the luminosity of our star.

In 1610, Galileo used one of the first telescopes to observe stars and planets, and discovered 'sunspots', dark regions on the surface of the Sun. He (wrongly) interpreted these as solar clouds, while today we know that they are actually slightly colder regions than the rest of the star's surface, which has an average temperature of just over 5,700 degrees Kelvin. Sunspots are characterized by a particular structure of the magnetic field and have a well-defined life cycle: they are born, they move towards the solar equator and they finally disappear. The number of spots present at any time varies in an irregular cycle of around 11 years, associated with variations in the intensity of the solar magnetic field. The latter actually inverts its polarity every 11 years, but the number of spots is linked to the intensity of the field regardless of its polarity. Figure 13.5 shows sunspots from Galileo's first observations to the present day.

As well as the 11-year cycle, the figure also shows a 'centennial scale' modulation in the number of sunspots, with minimums in the first decades of the nineteenth and twentieth centuries. Particularly evident is the period between 1650 and 1700, with an almost total absence of sunspots, known as the Maunder Minimum, named after the British astronomers, Annie Scott Dill Maunder and Edward Walter Maunder, who first proved the existence of this anomaly.

Sunspots garnered interest from a climate perspective because the total energy emitted by the Sun also varies with the same periodicity as the sunspots, and is greater when there is a larger number of sunspots, as shown by satellite measurements for the most recent cycles. From here, it is natural to associate the Maunder Minimum with a lower solar emission and, consequently, with the cold decades of the Little Ice Age, which could therefore have been generated by a variation in

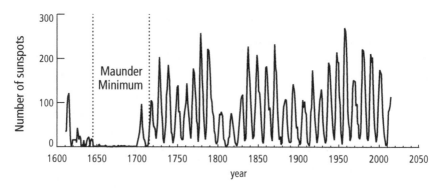

Figure 13.5 The number of sunspots (on a monthly average) from
Galileo's observations to the present day. The energy emitted by the
Sun varies slightly with the number of sunspots. *Source:* https://
solarscience.msfc.nasa.gov/SunspotCycle.shtml.

the energy received from the Sun. Similarly, it is sometimes suggested
that the Medieval Warm Period might have been caused by greater solar
emissions, of which, however, there is no clear evidence.

The purely 'solar' interpretation of these two climate fluctuations,
however, implies a substantially passive view of the Earth's climate and
ignores the processes of amplification and feedback that we have learned
about in the previous chapters. Furthermore, the variation of solar
luminosity in the 11-year cycle is around 0.1 per cent of the average value
of the solar constant, and therefore very small, as measured by satellites
over the last four solar cycles. We do not have direct measurements of
how solar luminosity varied during the Maunder Minimum, but if it was
to a similar degree, it was probably too low to trigger the Little Ice Age,
which in any case began well before the solar minimum. Furthermore,
there is ever greater evidence of localization of both the Medieval Warm
Period and the Little Ice Age in the region around the North Atlantic.
In short, the Sun does not seem to be the culprit, or at least not the only
one. If it did act, it must have had accomplices.

The Accomplices Within

In 2017, English solar physicist Mathew Owens and his team published
research that analysed the possible causes of the Little Ice Age, reaching

the conclusion that the variation in solar luminosity was only one of several causes. The others included a rise in volcanicity, with a release of dust into the upper atmosphere that probably reduced the quantity of solar energy absorbed by the surface. For example, in 1815 there was a devastating eruption of the Tambora volcano in the Sonda archipelago, which had such an extensive impact that the dust thrown into the atmosphere led to 1816 being an extremely cold year, especially in North America and Northern Europe. I would note how such a 'Year Without a Summer' (as it is now known) had severe repercussions on harvests, with serious famine as a consequence. Additionally, changes in the use of the land and vegetation could have played a significant role in the rise of the Little Ice Age.

Finally, we can call on an old friend: the thermohaline circulation. Wallace Broecker, whom we have already cited many times, suggested in 2000 that those fluctuations in the deep oceanic circulation on a millennial scale, responsible for the Dansgaard–Oeschger oscillations that were so evident in the ice age, could have continued in a reduced, weaker form during the interglacial period, leading to an alternation between warmer and colder periods and, in particular, the cold period of the Little Ice Age. The essentially European and North American local-ization of this cold period is effectively in keeping with the variability of the thermohaline circulation.

In any case, as we can clearly see from figure 13.2, from halfway through the nineteenth century there is a rapid rise in temperature on a global scale that extends throughout all continents. The speed of this rise in temperature is unusual and much more significant than anything seen in previous centuries. Something new was happening.

14

The Age of Humanity

The probe travelled through the darkness of the cosmos, having left a small planet, the third in its solar system, millennia ago. Its instruments had been switched off for a long time now and no more messages were sent back to its ancient creators. Before diving towards the infinite, the *Voyager 1* probe had taken another image of the place it was from, now reduced to a pale blue dot on a black background scattered with stars. Onboard it carried its precious cargo, a gold disc with a message thought up by a lively, uneasy species, capable of wonders and wickedness, with a desire to explore, to know, to act. A species who used that small amount of information in an attempt to tell their story, to say something of themselves to unknown – and perhaps non-existent – inhabitants of other worlds. Living beings who had been capable of modifying their planet for the better and for the worse, and who looked out towards the universe. Beings who knew how to learn from their own mistakes, correct their path, repair the damage they had caused. They were capable of this, though they did not always want to do it.

The Keeling Curve

In 1958, US geochemist Charles David Keeling of the Scripps Institution of Oceanography in San Diego began a long-term project to measure the concentration of carbon dioxide in the atmosphere through the analysis of the infrared radiation absorbed by the gases sampled. An observatory was established in Mauna Loa in Hawaii, far from direct sources of carbon dioxide linked, for example, to the presence of large cities or power plants. The research, financed by the US Department of Energy, allowed for the daily measurement of atmospheric CO_2 concentration, which continues today, accompanied over the years by new instruments installed in similar observatories located in remote areas throughout the world.

Figure 14.1 shows the curve of CO_2 concentration, rightly known today as the Keeling Curve, measured at Mauna Loa between March 1958 and October 2022. This trend is similar to what has been measured in all other observatories and reveals two important characteristics. The first is an annual oscillation with a variable range of between 7 and 10 ppm, with a minimum between August and October, observed and described by Keeling himself in a scientific publication in 1960. This annual variation is present in all of the signals measured by the various observatories, albeit with a slightly different range and with the minimum and maximum months varying slightly between the different latitudes.

But what causes this oscillation? Well, we could say that it is generated by respiration in the biosphere, and in particular by the seasonal cycle of vegetation in the Northern hemisphere. During photosynthesis, in the months of abundant growth the vegetation absorbs CO_2 in order to form new biomass, while in the autumn and winter months, when photosynthesis is scarce or absent, respiration dominates and so, therefore, does the emission of carbon dioxide. The CO_2 that is absorbed in the summer is therefore released, for the most part, in the winter.

An obvious cause for bewilderment at this point comes from the fact that, when it is summer in the Northern hemisphere, it is winter

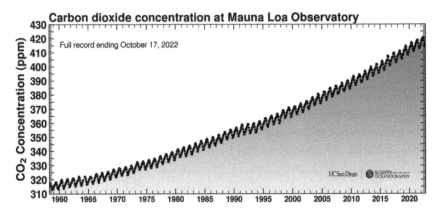

Figure 14.1 The Keeling Curve, which shows the concentration of carbon dioxide in the atmosphere in ppm, measured at the observatory of the Scripps Institution of Oceanography at Mauna Loa in Hawaii, from March 1958. Taken from https://keelingcurve.ucsd.edu.

in the south, and therefore the effects of the vegetation should cancel each other out. In reality, this is not the case because, given the current position of the continents, the majority of the landmasses are situated in the Northern hemisphere and therefore the contribution to the annual cycle from vegetation in the North dominates. The decrease in CO_2 concentration in spring–summer in the Northern hemisphere is partially compensated for by the simultaneous autumn–winter rise in the Southern hemisphere, but the difference is significant and the vegetation cycle remains visible in the carbon dioxide curve. The atmospheric transport processes and the details of which biospheric sources are more relevant in each specific observatory generate minor discrepancies and differences in the oscillations measured at different latitudes and positions, though for the most part the annual variation always remains visible.

An Explosive Growth

The second aspect of the Keeling Curve, which is absolutely dominant, is that the concentration of atmospheric carbon dioxide has grown very rapidly from just over 300 ppm at halfway through the last century to little less than 420 ppm at the beginning of 2022. This is a growth of around 100 ppm in 50 years, or around 2 ppm/year and there are no signs of it slowing down.

To clarify just how fast this growth is, figure 14.2 shows the concentration of carbon dioxide measured at Mauna Loa combined with that reconstructed from polar ice cores for the last 10,000 years. The change in the pace of CO_2 growth is astounding. If we think back to what we have learned about Antarctic ice cores, we can see that in the last 800,000 years, CO_2 concentration never went beyond 300 ppm. However, in the last century alone, the composition of the atmosphere has gone from what it had remained over the last hundreds of thousands of years to resembling much more closely the atmosphere of the times that preceded the last ice age.

As we have already done when discussing past climate variations, in this case too we can ask ourselves what gave rise to such a rapid increase in atmospheric CO_2. In these decades, there were, thankfully, no giant volcanic eruptions as happened in the Earth's past. Nor has there been

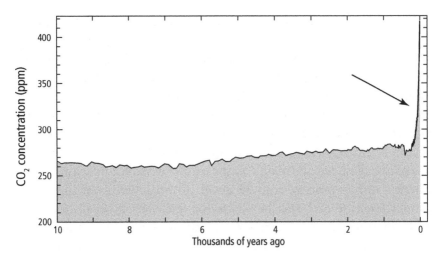

Figure 14.2 Concentration of atmospheric carbon dioxide in the last 10,000 years, reconstructed from ice cores and measured at Mauna Loa after 1958. Taken from https://keelingcurve.ucsd.edu. The data from the cores is from Dieter Lüthi et al. (2008) and C. MacFarling Meure et al. (2006), while the data from Mauna Loa is described by Charles Keeling et al. (2001).

a significant slowing of surface rock weathering, with the consequent reduced absorption of carbon dioxide. Nor has the biosphere's activity changed so drastically as to cause such a modification of the atmospheric composition. This time the answer lies in human activity. Since the Industrial Revolution, a new player has significantly modified the balance of atmospheric carbon dioxide. With the massive use of fossil fuels and the development of technologies that release huge amounts of CO_2 (such as the production of cement), or the emission of methane from both industrial and agricultural sources, humanity has started to influence the planetary carbon cycle, triggering a series of consequences for the climate.

The Carbon Cycle Again!

In previous chapters we have seen how the carbon cycle, one of the great global biogeochemical cycles, has two main branches. The slower geological cycle includes the release of greenhouse gases principally through volcanic and metamorphic processes, and the absorption of carbon dioxide through the weathering of surface rocks on the continents. The quantity of carbon in the crust and the mantle is formidable, yet the slow nature of the geological processes means that every year the total exchange of carbon between its main 'stores' (mantle, crust, atmosphere) is relatively modest. It is difficult to make a precise estimate but it is currently believed that, each year, a total quantity between 5,000 million and 1 billion tonnes of carbon is exchanged (emitted, absorbed and recycled). The book *Deep Carbon*, which collects contributions from various research groups, provides an up-to-date picture of how much we know today about the deep carbon cycle.

The biological cycle, however, is faster and, while it is linked to stores that hold a much smaller quantity of carbon compared to the mantle, it leads to exchanges that are estimated today to total around 200 billion tonnes of carbon a year. The biological cycle principally involves photosynthesis (both by plants on land and by marine algae and cyanobacteria) – which absorbs atmospheric carbon dioxide and releases oxygen – respiration and decomposition – which lead to a release of carbon dioxide (and methane for decomposition) from the soil and the biosphere into the atmosphere. Respiration is linked to the physiology of both plants and animals as part of their normal metabolism, and to microbial processes in the soil. Around half of the carbon exchanges are due to the functioning of terrestrial ecosystems, and the other half to marine life.

Over brief time periods (in a geological sense), the two cycles are roughly balanced, meaning more or less the same quantity of CO_2 enters and leaves the atmosphere. In certain periods, such as those of deglaciation, temporary imbalances can emerge, due for example to a greater extension of swamps or due to changes in the types of vegetation, which lead to oscillations in the quantity of atmospheric carbon dioxide, as we have seen when looking at glacial cycles. Roughly speaking, and in the absence of giant volcanic eruptions, variations in atmospheric CO_2 over periods of hundreds or a few thousand years tend to be of biological origin, while slower variations, those over tens or hundreds of thousands of years, probably include a significant contribution due to modifications in the geological carbon cycle.

In neither case in the past, however (at least as far as we can tell from glacial archives and marine sediment cores), has there been a variation in the concentration of carbon dioxide as fast as that seen over the last 100 years. Not even at the beginning of the PETM (discussed in chapter 8) or at the end of the glacial peak did atmospheric CO_2 vary so quickly. The estimate for carbon dioxide emissions of anthropic origin for 2014 is just under 10 billion tonnes, as we can see on the site of the US Environmental Protection Agency (EPA), and in 2018 it surpassed even this level. Detailed emissions estimates are also contained in the reports of the Intergovernmental Panel on Climate Change (IPCC). The IPCC, which was awarded the Nobel Peace Prize in 2007, plays a critical role in collecting the research published by scientists all over the world, which is then scrutinized and summarized in periodic reports that act as a source of essential information for understanding the changes that are currently taking place in our climate.

The Isotopes Return

Anthropic carbon dioxide emissions can be considered an important disruption of the geological carbon cycle, as they are caused in great part by the extraction of fossil carbon (oil, coal, etc.), which is burned and quickly released into the atmosphere. These are, however, also significant when compared with the biological cycle. It is true, in fact, that the latter exchanges greater quantities of carbon than anthropic emissions, but its cycle is balanced. In every store, more or less the same amount of carbon enters as leaves, and any eventual temporary imbalances are a very small fraction of the total amount exchanged. However, in the case of anthropic emissions, the flow only moves in one direction: from under the ground into the atmosphere. This is a clear displacement of carbon from a geological store into the atmosphere.

But how can we know that the rise in atmospheric CO_2 concentration is due to anthropic emissions and not, for example, a greater widespread emission through faults or some other geological mechanism? An initial, obvious response is that we have an estimate for the oil and coal consumed by many nations, and therefore of the quantity of carbon dioxide released into the atmosphere. In addition, we can find a clear response to this question in the isotopic composition of atmospheric carbon.

In the previous chapters, we have seen how photosynthesis tends to absorb more of the CO_2 containing the lighter carbon-12 than that containing carbon-13. The organic matter is therefore generally poorer in carbon-13 than the atmosphere, sea or rocks. And so, over recent decades, the ratio of carbon-13 to carbon-12 in atmospheric CO_2 has been diminishing in parallel with the rise in the total concentration of carbon dioxide. This tells us that there is more and more CO_2 of organic origin in the atmosphere. This is the precise origin of fossil fuels, because they were generated through the transformation of material that was once living and they therefore have a low isotopic ratio of carbon-13 to carbon-12. The great rise in atmospheric carbon dioxide, therefore, does not come from widespread CO_2 emissions of geological origin, but from anthropic emissions linked to the use of fossil fuels of ancient organic origin.

We also need to remember that not all the carbon dioxide emitted by human activities remains in the atmosphere. A significant part is absorbed by the oceans. Part is also absorbed by the vegetation, which, due to the 'fertilization' caused by a greater availability of CO_2 and higher temperatures that lengthen the growing season, tends to generate more biomass and stockpile carbon in its roots and in the soil whenever nutrients and water are present in sufficient quantities. But it is these higher temperatures that also favour more intense bacterial respiration, and so the resulting equilibrium between emission and absorption becomes a delicate one. Furthermore, tropical deforestation releases more carbon dioxide into the atmosphere, countering the CO_2 absorption generated by the increased vegetation activity. Summing up all of the contributions, the estimates (as reported by the American Meteorological Society, AMS) indicate that around half the carbon dioxide emitted since 1850 has remained in the atmosphere, leading to current concentrations.

The Time of the Anthropocene

Large-scale use of fossil fuels has generated significant changes in the composition of the Earth's atmosphere, taking carbon dioxide concentration to levels unseen in the last 800,000 years, and that have probably not been present for at least 3 million years. Our species has, then, been able to modify the planet's global environment, and not only through CO_2 emissions. This awareness led atmospheric chemist Paul Crutzen (awarded the Nobel Prize in Chemistry in 1995), at the beginning of this century, to revisit and use the term 'Anthropocene' to indicate the most recent geological period, characterized by a pervasive human (anthropic) influence over the entire planetary environment. The term was introduced in the 1980s by biologist Eugene Stoermer, with whom Crutzen published a discussion on the meaning of the Anthropocene in 2000. Even though the term has not yet been officially accepted into geological chronology, it has nevertheless become widely used – even if it is a little abused at times.

The idea of an era in which the anthropic effect has become dominant is not a new one. As early as the end of the nineteenth century, Italian geologist and paleontologist Antonio Stoppani proposed the expression 'Anthropozoic era' to indicate precisely that new and dominant human intervention. Vladimir Vernadsky, whom we met in chapter 6, also put forward a similar concept, pushing himself to consider the birth of the Noosphere, dominated by the human mind and knowledge.

But when did what we are calling the Anthropocene actually begin? The accepted idea is that it began with the Industrial Revolution in the first half of the nineteenth century, with the increasingly widespread use of coal and, later, other fossil fuels. So, it started when humanity effectively began to heavily modify the composition of the atmosphere.

However, US paleoclimatologist William Ruddiman suggested that the Anthropocene began more than 8,000 years ago with a human influence that started out slowly, before gradually becoming more evident. The modification of ecosystems, extensive deforestation to make way for agriculture, the emission of greenhouse gases from rice paddies: from this perspective, the period dominated by anthropic action substantially begins with agriculture and humanity's passage from the nomadic existence of hunter-gatherers to a community of farmers. According to Ruddiman, this slow-burning Anthropocene would have altered the normal oscillation between glacial and interglacial periods, at least slowing the end of the interglacial period in which we live and increasingly delaying the beginning of those processes that could lead to a new glaciation.

It has also been proposed that the Anthropocene truly began only after the Second World War, with the nuclear explosions in the atmosphere (whose radioactive fallout is still measurable today and will be for long to come, signalling a precise date in the sediment that will be studied by future geologists, whatever species they may be) and the impetuous development of industry and energy consumption throughout the world. In any case, the formal beginning of this era dominated by human action is, essentially, nothing more than a matter of definition. Much more important is the result of this action, in which the human capacity for modifying the environment has been deployed at maximum power.

Not Just Climate Change

Before ending this chapter, it is important to point out how the anthropic impact on the planet's environment is not characterized exclusively by the modification of the atmospheric concentration of carbon dioxide. Human activity has heavily impacted many other aspects in myriad different ways. Of primary importance is the rapid decrease in biodiversity, which is leading to a loss of living species before their existence is even discovered. In the last few years, the loss of biological diversity has been amplified to the extreme, to the point many are talking of a 'sixth extinction', this time caused by human activity. This extinction is the latest of the great biological catastrophes, generated in the past by volcanic mega-eruptions, sudden changes in climate or asteroid impact. The loss of biodiversity is, to all intents and purposes, one of the most worrying consequences of human actions.

Also fundamental is the change in land use, with the extensive deforestation carried out over past millennia to make way for farming and agricultural activity. This deforestation still continues unfettered in tropical regions and is motivated by global economic factors, whilst in other areas, such as in Europe, there is a tendency towards reforestation. In any case, the fragmentation of natural habitats has increased almost everywhere. The spatial continuity of the ecosystems is interrupted by infrastructure, roads and cities that reduce the extension of forests, prairies and wetlands, impeding the movement of living organisms and contributing to a decrease in biodiversity. Cities continue to grow, they absorb the land around them and host an ever greater percentage of people, even generating new anthropized ecosystems. Cities, which have been places of gathering, innovation and culture for millennia, now risk becoming disturbing megalopolises with no way out, enough to inspire those explored in the films *Blade Runner* and *Brazil*.

Pollution characterized the birth of the industrial era. Smog like London's famous 'pea souper' and fine particulate matter have long been inevitable bedfellows, unwelcome and dangerous for life in the cities and often the surrounding countryside. In the Western world, legislation on air quality passed in the 1980s greatly improved the situation, but it continues to be critical in many recently industrialized countries. Deaths from respiratory illnesses linked to pollution continue to claim victims all over the world. And water pollution in rivers, lakes and underground aquifers has reached extremely elevated levels. As well as organic and chemical pollution, there is the added danger of the plastics and microplastics that fill our planet's oceans and which have even been found in Alpine snow and in the most remote parts of the Arctic and Antarctic.

The use of nitrogen fertilizers, which allowed for the explosive development of agriculture, has led to an accumulation of nitrogen in the land and water that has upset the biogeochemical cycle of this element and led to the eutrophication of many aquatic environments. Currently, this aspect, together with the loss of biodiversity, is probably one of the most worrying and insidious.

Finally, there is the growth in the global human population, which at the time of writing is getting closer to a total of 8 billion people and shows no sign of stopping. Driven by the greater availability of food, more efficient medical treatment and longer life expectancies, this becomes a danger in itself, because it will require ever more resources, ever more energy, ever more land, especially if all the people of the world will want – rightly so – to reach a level of well-being that goes beyond pure survival.

Herein, then, lies the defining challenge of our times: to overcome the negative aspects of the inheritance of the Anthropocene, building instead on the positive ones, of which there are many. Will we manage? We must try, regardless. Otherwise, the gold disc on board *Voyager 1* we referenced at the beginning of this chapter might, in the future, tell the story of a species that no longer exists, and we do not want that. It will not be easy, but all we can do is commit ourselves to moving beyond all this. Garden cities, artificial photosynthesis, renewable energies with low greenhouse gas emissions, development that works in harmony with the laws of nature and not against them. Perhaps old Vernadsky and Teilhard de Chardin were not completely wrong to think in a visionary way of an era dominated by the creativity and knowledge that will come after the Anthropocene. Or at least to hope for such a possibility.

We will come back to this at the end of the next chapter. But, first, we are going to explore the changes to the climate over the last 100 years.

15

Global Warming

The quantity of information on how the climate has changed over the last 100 years and what the impact of this will be on our planet is now massive. Monitoring networks spread throughout the world, measurement campaigns, constellations of satellites dedicated to the observation of Earth that cover the entire globe – all provide a vast and articulated system of knowledge that is analysed and built upon by tens of thousands of researchers all over the world. In 1988, the UN and the World Meteorological Organization (WMO) created the Intergovernmental Panel on Climate Change (IPCC) composed of experienced scientists with a proven track record. The aim of the IPCC is to prepare periodic reports that collect and analyse the articles published by the scientific community, producing an up-to-date summary of how much we know about the climate. These reports (the three parts of the sixth were published in 2021–2) allow experts and interested bystanders to orient themselves within the vast landscape of climate studies. Here, we will discuss only the principal aspects of the global warming currently in action. Over the next few chapters, we will deal with some of the specific impacts and discuss what is being done to tackle their consequences.

One Degree in a Century

Over the last 100 years, the average global temperature on our planet's surface has risen by just over 1 °C, with an accelerated increase over the last 50 years. This increase has been on a global scale, and the second decade of this century has been the hottest of recent centuries. Years 2020 and 2016, in particular, set the record for the highest average global temperatures, a record that is in all probability destined to be surpassed in the near future.

Figure 15.1 shows the progress of average global temperatures observed by measurement networks distributed throughout the world and run by

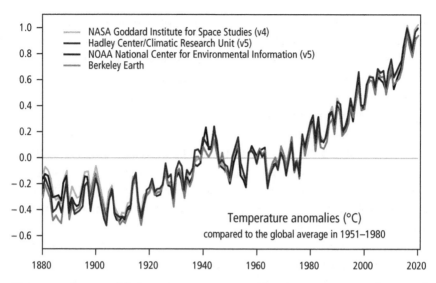

Figure 15.1 Average global temperatures measured by observation networks around the world. Here we can see the anomalies (differences) compared to the average in the period 1951–80. The curves, which are extremely similar, refer to different data elaboration carried out by the main centres for climate research. *Source:* NASA GISS / Gavin Schmidt, www.nasa.gov/sites/default/files/thumbnails/image/temp -2020_comparison-plot.jpg.

various centres for climate research, including the US agencies NASA (National Aeronautics and Space Administration) and NOAA (National Oceanic and Atmospheric Administration). Estimates at the time of writing, from the sixth IPCC report, give about 1.1 degrees for the observed global surface warming in 2011–20 compared to 1850–1900, and more than 1.5 degrees over land. In addition, the warming has accelerated since 1950, with no current signs of slowing.

The curve in the temperatures shown in figure 15.1 provides us with a number of important pieces of information that go beyond the obvious rise in average global temperatures. The first is the variability in average temperature from one year to the next, with the possibility that a very hot year is followed by a colder one, and vice versa. This is an essential characteristic of the climate and is called interannual variability. We cannot easily see it in paleoclimatic data, because we rarely manage to resolve fluctuations on an annual scale, but it is an intrinsic trait of the climate system. A slightly cooler year is not enough, therefore, for us to declare the end of global warming, as those in pursuit of an easy scoop often do. As always in science, it takes patience, data analysis and an accumulation of evidence and counter-evidence, which necessarily require time and careful study.

A second aspect is the fact global warming is not a smooth process, but one characterized by an alternation between periods of apparent stasis and others of frantic growth. For example, between 1945 and 1980, global temperatures did not rise, and also between 2000 and 2010 there was only a slight increase in temperatures, which was then compensated for by a violent increase after 2010.

There are many reasons for this behaviour and they tell us once again about the complexities of the climate system, which is capable of responding in a dynamic and irregular way even to a force that grows continuously, such as the atmospheric concentration of carbon dioxide. For example, the absorption of heat by the deep ocean is a process determined by ocean circulation and the mixing of the surface water layers, and it is possible that this caused the period of limited temperature increases in the first ten years of this century. In 2013, climatologist Yu Kosaka, who is now in Japan, and oceanographer Shang-Ping Xie of the Scripps Institution of Oceanography suggested that the 'hiatus' of the early 2000s was caused by a modification of the currents in the equatorial Pacific, associated with a series of intense *La Niña* events and a temporary strengthening of equatorial atmospheric circulation. These mechanisms presumably shifted part of the surface heat to the deepest marine layers, slowing global warming, which, however, returned with renewed vigour in the following years.

The stasis in temperature immediately after the Second World War, however, could have been caused by a huge release of aerosols (atmospheric dust particles) associated with industrial production. Put simply, there are two types of aerosols: those that are carbonaceous and dark, like the soot produced through combustion, and those that are sulphur-rich, light, generated by chemical reactions of sulphuric acid and produced in large quantities by various industrial processes. Both are dangerous to our health, but they have different effects on the climate: soot absorbs solar radiation and heats the atmosphere, cooling the surface below it, whilst sulphate aerosols reflect sunlight and cool both the atmosphere and the surface.

Following the Second World War, industrial production rose at an accelerated rate throughout the West, leading to the emission of large quantities of sulphate aerosols. So it might be that these very particles slowed the warming caused by the simultaneous emissions of carbon dioxide. In the 1980s, new policies were put in place regarding the emission of atmospheric particulate (as aerosols are commonly known) with more severe regulations over fossil fuels, industrial plants and pollution generated by means of transport. The effect on health was enormously positive, but solar radiation, no longer reflected by these particles, began once more to add to the effects of CO_2, and global warming accelerated sharply.

In the curve seen in figure 15.1, therefore, we must distinguish between two aspects: the average growth of around 1 degree in 100 years, and fluctuations over a few decades caused by the many variability mechanisms of the climate system. The climate is lively, rich in unexpected behaviour, uneven. Not only over time, but also in space.

The Climate's Hotspots

Global warming should not be thought of as a kindly uniform rise in temperature throughout the planet. It occurs in sudden leaps, as we have just seen, and it is not the same in all areas of the world. Figure 15.2 shows the differences in average temperature in the months December–January–February and June–July–August in the 30-year periods 1981–2010 and 1951–80, obtained by the data analysis carried out by climatologist Marco Turco (now at the University of Murcia in Spain) when he worked as part of our research group at the Consiglio Nazionale delle Ricerche (CNR) in Italy.

In most of the planet, the temperature has risen significantly, but the warming has been particularly intense in certain areas. For the most part, in continental areas at high latitudes in the Northern hemisphere, such as Siberia and the most northern parts of the American continent, the difference in winter temperature between the two periods considered has reached, and sometimes surpassed, 2 °C.

As well as the temperature, precipitation (in particular, its distribution over space and time) is of primary importance to the environment,

DJF

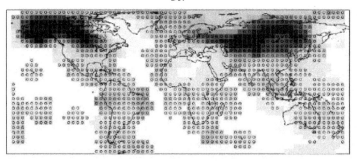

JJA

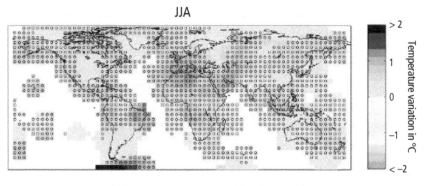

Figure 15.2 Average difference in temperature for the months December–January–
February (DJF, at the top) and June–July–August (JJA, at the bottom) between
the 30-year periods of 1981–2010 and 1951–80 for the continental areas. Taken
from Marco Turco et al. (2015). The colour originals are freely available on the site
https://agupubs.onlinelibrary.wiley.com/doi/full/10.1002/2015GL063891.

the ecosystems, agriculture and human societies. But how has global
precipitation changed over the last 100 years? The total quantity of
annual precipitation, averaged out throughout the globe, shows strong
variations from one year to the next, and slower fluctuations over the
decades, but the overall increase is very modest. To simplify, we could say
that, on a global scale, average precipitation has not changed significantly
over the last century. However, what has changed – and in some cases
drastically so – is the distribution of precipitation.

Figure 15.3 shows the percentage difference between precipitation
in the 30-year period of 1981–2010 and that of the period from 1951 to
1980, once again analysed by our research group. The results show how,

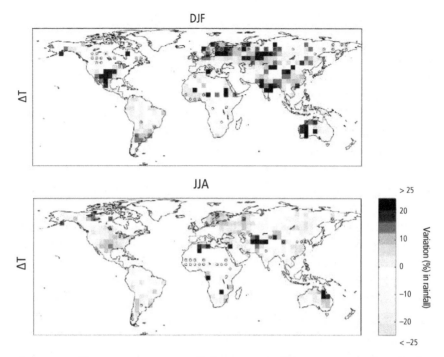

Figure 15.3 The average percentage difference in total precipitation in the months
December–January–February (DJF, at the top) and June–July–August (JJA,
at the bottom) for continental areas between the 30-year periods of 1981–2020
and 1951–80. Taken from Marco Turco et al. (2015). The colour originals
are freely available on the site https://agupubs.onlinelibrary.wiley.com/doi/
full/10.1002/2015GL063891.

over the last 30 years, some regions of the world have become decidedly
drier than during the 30 years before, whilst others have become rainier.
For example, all of Northern Europe has seen a significant rise in
precipitation (and flood risks), while Mediterranean Europe has been
characterized by a significant drop in rainfall, as is discussed in much
greater detail by the European Environment Agency. This decrease,
along with the increase in temperatures, means a rise in drought
conditions and greater pressure on water resources, vegetation and
agroecosystems.

Another region that has been hit particularly hard by changes in
precipitation is the Sahel, south-west of the Sahara, where the already

scarce rainfall has decreased even further, causing drought conditions that are extremely problematic for local populations, generating famine and humanitarian catastrophes, as happened in 2012 with consequences that still resonate today.

In 2006, climatologist Filippo Giorgi, who now works at the Abdus Salam International Centre for Theoretical Physics in Italy, coined the phrase 'climate change hotspots' to indicate those areas of the world in which the change has been most intense. It is necessary, however, to define what is meant by 'intense'. The path chosen by Giorgi (not the only one, of course) is that of using a certain number of indicators such as variations in temperature and precipitation and their interannual variability, the change in the number of extremely hot days or of those that are extremely dry or extremely rainy. By following this path, we identified the 'hotspots' using the data shown in the two previous figures. The critical regions that emerged from our analysis are, essentially, the Sahel, tropical western Africa, Amazonia, Indonesia and central–eastern Asia, but also the Arctic and, albeit to a lesser degree, the Mediterranean basin. These are, therefore, under special surveillance, because it is in precisely these locations that the combined variations of different climate parameters are particularly important and can cause effects that amplify one another in a reciprocal manner.

Raining Cats and Dogs

As well as changes in the geographical distribution of precipitation, there are numerous indications that rainfall intensity is also varying, as we might expect given what we have learned about the water cycle in periods with a very hot climate (in the Eocene, for example, described in chapter 8). Today, in various regions, we can observe a higher concentration of precipitation, with more intense rainfall interspersed with dry periods that last longer than in previous decades.

But how does this intensification process work? In chapter 9, we saw that a hotter atmosphere can hold a greater quantity of water vapour. If there are sufficient water reservoirs on the planet's surface that can evaporate, we can then expect that with a rise in temperatures the amount of vapour in the atmosphere also grows. This also means that there will be more energy available for atmospheric motions and, ultimately, a possible intensification of the water cycle. But the exact dynamics we are heading towards are not yet clear.

Measurements taken over recent decades indicate that, in the case of Earth, both the intensity of precipitation (how much it rains during precipitation events) and the average duration of the periods without precipitation have increased. Filippo Giorgi introduced an index he named HY-INT (hydrological intensity), defined by the product of the average intensity of precipitation during an event and the typical duration of periods without rain, demonstrating how this index is rising in many regions of the world. This means the water cycle is changing, with an increase in both the violence of single precipitation events and the duration of dry periods.

Similar results were obtained by a research group at the United States Geological Survey (USGS), led by hydrologist Thomas G. Huntington, who demonstrated how the water cycle is becoming more intense throughout most of the United States, due first and foremost to changes in the precipitation regime. A 2019 study by hydroclimatologists Daniel Bishop, Park Williams and Richard Seager shows how, in the south-east US, autumn precipitation has increased over the last 120 years (as we can also see from figure 15.3), due essentially to a greater intensity in precipitation events. Elsewhere, however, the signs are not so clear, and the entire issue of changes in the water cycle remains one of the great conundrums faced by the scientific community studying the climate. We certainly cannot study this problem using a global average, because it is crucial to separate what happens on land from what happens at sea, and to distinguish between the different regions of the world, precisely because of the wildly uneven nature of all the variables associated with the water cycle.

A better characterization of the water cycle is, nevertheless, essential if we are to effectively tackle the negative impacts of climate change – even more so than for the rise in temperature. The presence of the same amount of annual rainfall concentrated in a lesser number of much more powerful events can lead to situations of great difficulty for people, their homes, infrastructure and agriculture. Extreme precipitation events, together with occupation of land by humans in a way that is too often unregulated, can lead to catastrophic flooding, landslides and mudslides, as we are unfortunately seeing with greater regularity in our daily news, and which are often followed by epidemics, famine, forced migration, migrant camps, loss of human life, and permanent economic and social damage.

Of course, it is difficult for us to distinguish here between a rise in damage caused by a real intensification of extreme events and that caused by a rise in exposure, with homes, settlements and infrastructure located in riverbeds or at the base of slopes that are known to be unstable. Sometimes, it is a case of carelessness, the pursuit of easy economic gain, ignorance or corruption. Other times, these are the only places where there is available land – take it or leave it. A great deal has been written on this crucial issue, and on hydrological instability and its consequences in general.

Ever Drier Summers

Equally important is the rise in drought conditions in many areas of the planet, such as in the Mediterranean basin, the west of the North American continent or the aforementioned Sahel, with potentially devastating consequences for the availability of water resources of sufficient quantity and quality and agricultural production, not to mention the prospect of possible wars over water.

Put simply, the hydrological balance in a portion of the soil is determined by the equilibrium between how much water comes from precipitation, melting snow and surface downflow from higher areas, and the water losses due to surface runoff to lower-lying areas, infiltration to surrounding aquifers and evapotranspiration. Evapotranspiration, which we encountered in chapter 6, is the sum of evaporation from the soil and transpiration from plants, which release water into the atmosphere through the stomata in their leaves.

The decrease in precipitation and the rise in temperature reduces the water supply and amplifies evapotranspiration, favouring conditions of lesser soil humidity and, ultimately, a greater likelihood of drought conditions. The European Environment Agency has a map of the variation in the average humidity of Europe's soil in the period from 1951 to 2021, obtained using a hydrological model that calculates the balance between precipitation and evapotranspiration. In southern France, northern Spain, Greece, the northern Balkan peninsula and in central-northern Italy, a decrease in the soil's water content has been estimated at more than 5 litres per metre every 10 years, consistent with the decrease in precipitation throughout the Mediterranean basin.

In the west of the United States, the last 20 years have been characterized by conditions that are among the driest ever observed over the last century, as proven by the analysis of NASA's satellite data and discussed in the work cited above by Park Williams and his colleagues. The effects of this drought will be long-lasting, with a significant impact on agriculture, farming and the economy in general, not to mention the lives of the people who inhabit that region.

The first part of the IPCC's latest report, published in August 2021, collated scientific work that analysed the increase in drought in different areas of the planet, condensing it into an impressive image (see figure 15.4) showing the variation of drought conditions from 1950 to today.

Almost everywhere that there has been a change, the incidence of drought has risen. And in two regions – the Mediterranean and the western United States – there is a high probability that the rise in drought has a direct link to global warming.

The alteration in the water balance also brings the possibility of a reduced recharge of the aquifers on which we depend for drinking water, agriculture and, partially, industrial use. Insufficient recharge means the aquifer's water level is reduced, often exacerbated by a more intense use of this water due to drought conditions and summer heat. In coastal regions, these changes favour the infiltration of marine water, which causes the salinization of the aquifers. Today, the quantitative understanding of how climate change and excessive exploitation are modifying the aquifers is an issue of critical importance, both scientifically and practically, if we are to plan better strategies for managing water resources. In short, water – too much or too little – is a difficult subject in which climate change, land use, social, economic and political aspects, legal issues and infrastructure planning all come together. It is a difficult yet essential issue for our future, one which must remain at the centre of our attention.

Fire, Fire!

The rise in temperature over recent decades has, in a great many regions of our planet, also led to an increase in heatwaves, as detailed in the

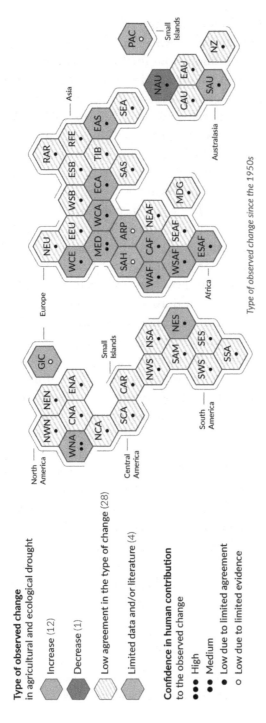

Figure 15.4 The variation of drought conditions throughout the world, from 1950 to today. In most of the areas where there has been a change, drought has increased. The western United States and the Mediterranean emerge as regions that are particularly affected by the increase in drought conditions. Taken from the IPCC's sixth report, *The Physical Science Basis*, 2021.

IPCC's sixth report. The example of Europe in 2003, or that of western Canada and Siberia in 2021, are just a few of the many episodes of this kind that we must record. Also associated with these heatwaves and drought conditions are the enormous forest fires that too often develop as a result, as we tragically saw in Canada in the summer of 2021.

Fire is a natural element of many ecosystems (the Mediterranean, for example) and the plants in these regions have adapted to its presence. A burnt forest regenerates over the course of decades, but in recent times we have built houses, roads and infrastructure almost everywhere and the fires have become a much more real threat, with huge loss of life and material goods. Almost every summer, dramatic episodes of large forest fires occupy the front pages. As well as those in the Mediterranean, the devastating fires in California and Australia speak to climate conditions that are increasingly favourable for fires. In the Mediterranean, where almost all the fires today are of direct anthropic origin, due to either carelessness or malice, the total area that burns every summer depends on the climate conditions, which can determine the speed with which the fire spreads and how extensive it is.

In other contexts, such as in Canada and at high northern latitudes, recent years have seen the development of vast fires even in regions with ecosystems that are not suited to fire. The effect in these cases is even more devastating and can have a long-term impact on the environment and the ecosystems that is difficult to foresee and control. For example, large-scale fires emit vast amounts of carbon dioxide and particulate (soot) into the atmosphere, which can amplify the effects of global warming. The European Union's Copernicus Atmosphere Monitoring Service (CAMS) has estimated that globally, in 2021, fires emitted around 6,540 million tonnes of carbon dioxide, a level that is circa two and a half times that of the EU's total emissions in 2020. These fires are observed in North America, Siberia, the central and eastern Mediterranean, North Africa, Australia and India. This is a global problem that requires a global solution, also because there are now more and more indications of the widespread occurrence of massive mega-fires capable of devastating extremely large areas.

But let's try to understand better how these fires occur. In order for a forest fire to develop, certain conditions are required. Anyone with access to an open fireplace knows that, first and foremost, there must be enough fuel in the form of dry sticks and flammable material, and that the firewood must not be wet. Climatologist Marco Turco and his colleagues developed a quantitative, empirical model of the link between summer climate conditions and the area that is burned, making it possible to obtain an educated estimate of the entire area involved in the fires, starting with the knowledge (or prediction) of summer temperatures and precipitation. This link, originally developed for Mediterranean ecosystems, is now being extended to other kinds of environments.

In any case, the example of the Mediterranean is important for understanding what happens and how we can tackle the problem. As summers become increasingly dry, we might expect a parallel increase in the burned area. However, the work carried out by Turco and his colleagues demonstrates how in most of Mediterranean Europe over recent decades, the area burned by summer fires has decreased. Figure 15.5 shows the record of the area burned by summer fires, using the data from the EFFIS (European Forest Fire Inventory System) archive and various national registers on a geographical scale known as NUTS3 (Nomenclature of Territorial Units for Statistics, level 3), which roughly corresponds to provinces or aggregates of provinces situated close together.

The fact that the burned area has decreased is obviously good news (the anomalous behaviour in certain areas is less satisfying), but it is important to understand the causes of this trend. The main one was the marked increase in fire prevention and control measures, with improved, real-time monitoring techniques that include the use of satellites, and rapid intervention from teams on the ground and in aeroplanes and helicopters by the civil protection services of various countries. Following the large Mediterranean forest fires in the 1990s, a great effort was undertaken to deal with this problem, demonstrating once again that, when faced with problems, including those caused by human activity, the only sensible response is to get to work repairing the damage inflicted.

Unfortunately, however, the situation is not as rosy as it might seem and we cannot be content with what we have done this far. First of all, there are still regions where the burned area has grown, and significantly so. Second, as we have seen, fires now tend to develop in areas that were not previously affected by fire and in seasons that go well beyond the summer. Finally, the predicted rise in drought conditions means that over the next few decades there will be a potential growth in the affected

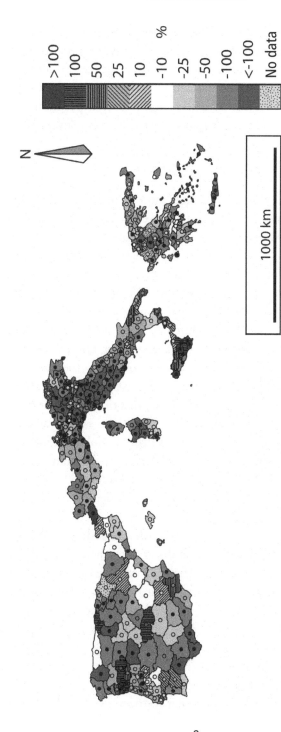

Figure 15.5 Variations by percentage of the area burned by summer fires in the period 1985–2011, obtained by dividing the total variation by the average area burned in the same timeframe. The data is taken from the European EFFIS archive and refers to geographical units known as NUTS3. This figure is taken from the work by Marco Turco et al. (2016). The colour version is freely available on the site https://journals.plos.org/plosone/article?id=10.1371/journal.pone.0150663.

areas, with a trend that could nullify the positive effect of the measures taken this far. We need, therefore, to improve further our prevention and control techniques, to refine our forecasting capacity on a seasonal level, and foster ever greater communication between the civil protection forces of various countries within a large international programme for controlling forest fires without getting distracted by less relevant issues.

Higher and Higher

Since 1880, sea level has risen, on average, more than 20 centimetres and at a faster rate in the last 25 years. In this case also, there are significant regional differences between the various basins caused by the character-istics of marine currents, winds and the dynamic response of the oceans.

Localized episodes of subsidence aside (i.e., the sinking of certain coastal areas, such as the Venice Lagoon), the rise in sea level is linked to the rise in temperature. It is estimated that somewhere between a quarter and half of the rise in sea level is generated by water having expanded for purely thermal reasons due to the temperatures being higher. This effect has become more significant in recent decades owing to the more rapid warming. To understand this mechanism, we must remember that water has a maximum density at around 4 °C. Above this threshold, the density of water decreases as the temperature rises, and the same water mass must therefore occupy an increased space. A rise in global temperatures of around 1 °C will therefore lead to an increase in the volume of the ocean waters, and, given that their 'container' (the total size of the ocean basins) is substantially fixed (on the time scales of decades at least), the sea cannot do anything but rise.

The remaining part of the increase in sea levels is due for the most part to the melting of continental ice sheets and glaciers, both polar and mountain. As happened during the Last Glacial Maximum, when the sea was lower due to large amounts of water being trapped in glaciers, today's speedy retreat of continental ice is releasing new water into the ocean basins, increasing both their mass and volume. In contrast, the melting of sea ice (in the Arctic, for example) does not actually alter the sea level. Sea ice is less dense than water and floats on the surface, partially submerged and partially emerging from the sea (like an iceberg). The water released by the melting of that ice occupies exactly the same volume as the portion that was submerged and does not, therefore, modify the sea level.

The rising sea levels pose enormous problems for coastal settlements and infrastructure. Think of Holland and its 'Netherlands', or the many inhabited islands in the Pacific Ocean. But all over, in Europe, America and Asia, there are enormous cities and industrial plants on the coasts that tend to be the most densely populated and activity-rich areas. The cost of moving these settlements, of the forced migration of coastal inhabitants or the construction of protective infrastructure could be enormous, and, for some nations, unfeasible. Furthermore, the rise in sea levels, often accompanied by more violent atmospheric perturbations due to the intensification of the hydrological cycle, carries a greater risk of coastal flooding and destructive waves, as we are already seeing, alongside an accelerated erosion of the coasts. Our planet's climate is dynamic, but a complex society such as our own, rich in fixed infrastructure, would find it very difficult to adapt to such rapid changes.

The Causes of Warming

The underlying question is: why have temperatures risen by approximately 1 °C in the last 100 years and, most importantly, why so quickly? Solar luminosity has not risen, nor are orbital variations able to happen in such a short time. As we already discussed in the previous chapter, there have been no volcanic mega-eruptions, the biosphere has not drastically modified the way it functions and nor have we seen dramatic changes in cloud cover.

Instead, over the last century, the concentration of carbon dioxide in the atmosphere has risen significantly, from less than 300 ppm to almost 420 ppm in 2021, and it continues to grow. As a result, as we saw in chapter 14, the greenhouse effect is amplified. Just enough to take global temperatures 1 degree higher than they were a century ago. Not by much, compared to the climate changes that have happened in the past, but very quickly, just like the concentration of carbon dioxide caused by emissions of anthropic origin.

Thousands of studies, research projects, measurements, data analysis and modelling studies (we will come back to climate models in the next chapter) have shown that the rise in temperature we are witnessing today has been generated for the most part by greenhouse gases released into the atmosphere by anthropic activity. We can confirm, therefore, with sufficient recognition of the cause, that a large part of the rise in temperature over the last century is generated by human activity. We too are engineers of the ecosystem, just like beavers and bacteria, and we act on a global scale. Though not always with sufficient awareness of the consequences.

In recent years, the results of research into the so-called attribution of the causes of climate change have often been presented in tables that show estimates of how much can be ascribed to one mechanism or another. These are clearly estimates, they are not easy to produce and are affected by uncertainty, which is in turn evaluated. Figure 15.6, taken from the IPCC's sixth report (AR6), shows a simplified estimate of the causes of the rise in temperature observed in the decade from 2010 to 2019, compared with the period 1850–1900.

Anthropic emissions of carbon dioxide, methane, nitrous oxide (N_2O) and other greenhouse composites have caused the surface temperature to rise by around 1.5 °C. Other environmental changes caused by human activity have, however, reduced this increase. Here, an important role has been played by changes in land use, for example with tropical deforestation and the subsequent rise in the reflection of light (albedo) from the bare soil, or with the emission of sulphate aerosols. These other effects are more difficult to gauge, as we can see from the larger uncertainty bars. For example, dark aerosols (soot or black carbon) have a warming effect on the atmosphere and, when they are deposited on snow, they diminish the albedo and accelerate its melting through a greater absorption of solar energy. The average rise in temperature is joined also by the natural variability of the climate, which continues to be present, for example with oscillations on a multi-decadal scale that modify and complicate the signal generated by anthropic activity.

Naturally, the entire climate system responds in an articulated way to the rise in the concentration of greenhouse gases. Part of the carbon dioxide has, as we have already said, been absorbed by the oceans, increasing their acidity. Part of the heat accumulating in the ocean surface layers has also been transferred to the deep waters, heating them slightly and slowing the increase of surface temperatures. The response by cloud cover, vegetation and the rise in water vapour contained in the

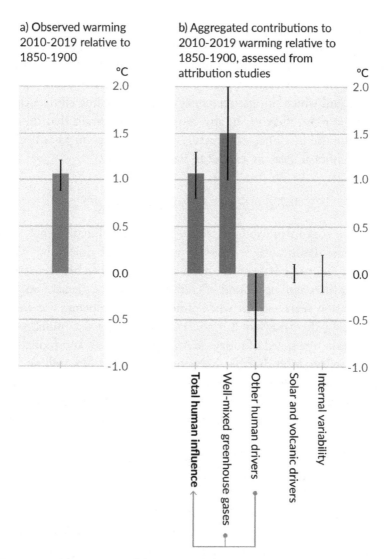

Figure 15.6 The estimate of the causes of climate change over recent decades. On the left, the warming that has been observed, as indicated by the difference between the average of the decade 2010–19 and that of the period 1850–1900. On the right is the estimate for the causes of this warming, the other forces of anthropic origin (particularly aerosols and changes in land use), natural forces (variations in solar luminosity and volcanoes) and the climate's natural variability. The thin bars indicate the range of uncertainty. Taken from IPCC AR6, Summary for Policy Makers.

atmosphere are all effects that can intensify or slow the rise in temperature depending on the context in which they are manifested. These effects, which are not yet fully understood, should be carefully studied in order to obtain even more precise estimates of what we can expect in the future and which mitigation strategies for containing climate change risks are the most effective. In any case, it is now certain that the main cause of the rise in global temperatures in the last century has been the rise in greenhouse gases as a result of human activity.

Who's Afraid of Global Warming?

As we have seen in previous chapters, the Earth's climate is not static. It fluctuates, significantly even, over the course of time. The average global temperature has been as much as 10 °C higher between 50 and 60 million years ago, or around 5 °C lower during the glacial maximum around 20,000 years ago. Or when almost the entire planet was covered in ice during the Snowball episode that took place 650 million years ago. However, the speed of today's warming is unusual, and probably far greater than anything that has happened in the last few million years. Even the increase in temperature at the beginning of the PETM or during the last deglaciation was significantly slower than it is now. In any case, even if we were to push the temperature to rise another 3 or 4 °C with our carbon dioxide emissions, we would not put the survival of our planet or its biosphere at risk. But our species and the societies in which we live today would perhaps not be so lucky.

Estimates indicate that, with an increase of 2 °C compared to the pre-industrial era (or about 1° more than what it is today), the economic and social costs as a result of the warming, as well as the loss of human life, would be much higher than the costs (though significant) of a conversion of our industrial plants and energy production. The rising sea levels, prolonged droughts, increasingly violent storms and damage to harvests could trigger uncontrollable mass migrations. Many low-income countries would not be able to cope with the difficulties, and the very fabric of human societies could be at risk, with a greater probability of war and a possible transition towards increasingly authoritarian societies. This is not a good world for our descendants to inherit, nor is it a good world in which to spend the final years of our own lives.

Faced with a serious problem, there are two attitudes that should be avoided. The first is to deny the existence of the problem, insisting that global warming doesn't exist or that it is not due to a rise in emissions of human origin. As well as being scientifically wrong, this attitude is dangerous as it distracts attention away from the real and difficult discussion on how to tackle the situation and which development models to focus on. The second wrong attitude is catastrophism, often accompanied by an implicitly negative view of modern life and a vague aversion to human activity, seen as always damaging. Sometimes, this way of thinking borders on a desire for a primal Arcadia, in which we lived in contact with nature and in harmony with the universe, though it is also where we tended to die at 30, our bodies consumed by fatigue and hardships, if we were lucky enough to survive childhood.

In reality, Arcadia has never existed. Human beings have always acted for their own well-being (individual or collective), at times successfully, at others poorly. The uncontrolled growth of the population and the unbridled consumption of resources with no respect for the planet's rules have generated serious and real problems, such as climate change, pollution, loss of biodiversity and fertile soil, and inequality between peoples. The problem exists. Denying this does not help. The only solution we have is to face it using all of our knowledge, our technology, our intelligence and our desire for justice. By respecting the rules of our planet, not for some abstract ideological reason but because they are unavoidable rules that allow for our survival.

16

Arctic Sentinels

Over the last 50 years, average temperatures in the Arctic have risen by around 2 °C. Since the beginning of the century, the speed of warming in the regions around the North Pole has been at least double that seen in the rest of the world. The Arctic is, therefore, an area that is severely affected by climate change and the current warming is modifying the environment at high latitudes in many different ways. The hyperborean kingdom of ice and snow is rapidly changing, making way for expanses of marshland and species coming from lower latitudes. Furthermore, what happens in the Arctic does not stay in the Arctic but influences the climate throughout the rest of the Northern hemisphere, and potentially that of the entire planet.

Arctic Amplification

An obvious question is that of why the Arctic is heating up so quickly compared to the rest of the planet, in what is defined as the 'Arctic amplification' of global warming. There are many reasons and, again, each one tells us more about the way the climate works, as US geographers and polar climatologists Mark Serreze and Roger Barry summarized in their article of 2011.

Firstly, the melting of land and sea ice, when it uncovers a new surface of soil or sea, reduces the albedo and allows a greater absorption of solar radiation by the surface, therefore accelerating warming. Perhaps even more important is the early melting of winter snow, which uncovers the darker land increasingly earlier in the spring and makes a significant contribution to the reduction of the albedo.

In addition, the reduction in sea ice leads to greater direct contact between ocean waters and the atmosphere, facilitating the transfer of heat and causing the humidity of the Arctic air to increase. This can lead to richer atmospheric dynamics, with more intense convective episodes and a different amount and type of clouds, as we already saw with the same (though much more powerful) Arctic warming during the Eocene. Even the changes in heat transport from low latitudes can play a significant role.

Finally, we have soot, including that generated by fires, which is often carried towards the Arctic by atmospheric circulation and ends its journey by being deposited onto the snow and ice, accelerating the rate at which they melt. All of these processes combined conspire to make the Arctic ever hotter.

It is interesting at this point to compare this with what is happening in the Antarctic, at the other end of the world. Following an initial period in which Antarctic sea ice increased, presumably due to an intensification of circumpolar winds that facilitated evaporation and, therefore, cooled the surface of the Antarctic Ocean waters, the situation has turned on its head. The oscillations in the extent of Antarctic sea ice from one year to the next became increasingly intense, until in 2014 there was an extremely rapid decrease – even more pronounced than in the Arctic – as demonstrated by NASA climatologist Claire Parkinson in her analysis of satellite data from the last 40 years.

The great Antarctic glaciers are not in much better health. They are more extensive (and, as we have seen, older) than those in the Northern hemisphere, and, so far, they have been more stable, particularly the East Antarctic ice sheet, which is the larger of the two. But the Western Antarctic Ice Sheet does not appear to be as robust, particularly as it is very sensitive to the temperatures of coastal ocean waters and has lost a significant amount of ice over the last two decades. The crucial challenge is to understand how the East Antarctic ice sheet will react: whether it will remain stable or whether it too will start to respond to marine warming. If it were to melt even partially, the East Antarctic ice sheet could cause sea levels to rise by several metres. So, even in Antarctica, there is no climate peace.

But let's return to the Arctic and see how this warming is affecting the overall environment of the higher latitudes.

Glaciers in Retreat

One of the most obvious effects of this warming is the drastic retreat of the glaciers and ice sheets. In many areas, the retreat is so evident that, every year, researchers need to go farther into the mountains in order to reach the glaciers. Figure 16.1 shows the example of the Bayelva river basin, which flows in Kongsfjorden on the island of Spitzbergen in Svalbard, Norway. The glaciers, visible in the background, retreat more every year, and the river carries large quantities of sediment to the sea. This area, as we will see shortly, is a research focus for various groups as part of an interdisciplinary effort to understand the changes in the Arctic environment in response to global warming.

For these reasons, the dynamics and mass balance of Arctic glaciers are studied by glaciologists all over the world. In Greenland, for example, the vast international collaboration of polar scientists called the Ice Sheet Mass Balance Inter-Comparison Exercise (IMBIE), sponsored by the ESA and NASA, and coordinated by glaciologist and expert in Earth observations Andrew Shepherd from Leeds University in England, has estimated the ice sheet's loss of mass using satellite data. The results indicate a negative balance – meaning summer melting surpasses winter accumulation – which in 2011 reached an extreme level with an annual loss of around 345 billion tonnes of ice. Overall, the estimate indicates a loss of almost 4,000 billion tonnes of ice between 1992 and 2018. In this

Figure 16.1 The basin of the Bayelva river on the island of Spitzbergen. In the summer, the river fills with sediment that is then released into the sea. The glaciers visible in the background, including the Vestre and the Austre Brøggerbreen, are retreating more each year. Photo by the author (Summer 2018).

period, the reduction of the Greenland ice sheet alone has contributed to a rise in global sea levels of around 1 centimetre.

There are various ways in which a large ice sheet or Arctic glacier can lose mass. For example, they can melt on the surface, generating lakes on the surface of the ice and a more intense flow of the meltwater, thus modifying both the surface and the internal hydrology of the glacier. Inside the glacier's mass, the water released by melting creates a complex network of channels and cavities that are continually changing. Part of this water can increase the liquid layer that forms between the ice and the sediment (or rock), accelerating the flow of the glacial mass. In Greenland, the speed of the glaciers' movement has increased, carrying the ice towards the ocean much faster, where it fragments into icebergs.

The research carried out by the Italian-born US oceanographer Fiamma Straneo has revealed a complicated interaction between the ice and marine circulation in the fjords. In addition to the local increase in temperature, marine circulation is bringing increasingly warm ocean water to the fjords, which then comes into contact with the front of the glaciers and accelerates the rate at which they melt. This, with its amplifying effect, in turn modifies and accelerates circulation in the fjord and the arrival of ocean water. Many of the measurements have been made (not without difficulty) in the Sermilik fjord in eastern Greenland, and later this mechanism was reproduced in the laboratory, with experiments carried out in tanks by oceanographer Claudia Cenedese, also of Italian descent, who has worked for many years at the Woods Hole Oceanographic Institution in the US. In 2015, Straneo and Cenedese published a review article in which they explained in detail the processes of interaction between glaciers and marine circulation, which is crucial when estimating the response of the great Arctic glaciers to global warming.

New Routes in Arctic Seas

Alongside the glaciers on land, the sea ice is also rapidly reducing. Figure 16.2 shows the extent of Arctic sea ice in September, the month in which it reaches its annual minimum, from 1979 to 2021. The data comes from measurements taken by NASA, the NOAA and the National Snow and Ice Data Center (NSIDC) in the US. As well as spatial distribution, the

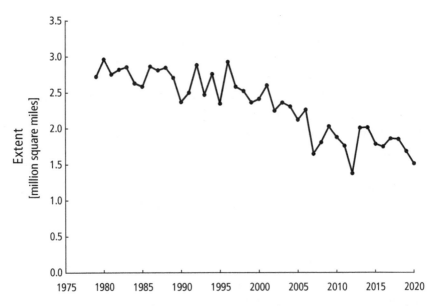

Figure 16.2 Average September Arctic sea ice extent from 1979 to 2021, using data provided by NASA, NOAA and NSIDC. Taken from www.globalchange.gov/browse/indicators/arctic-sea-ice-extent.

average thickness and, therefore, the total volume of Arctic sea ice has also decreased.

The decrease is impressive and has significant consequences for the climate, for Arctic ecosystems and for geopolitics. From a climate perspective, as we have said, the reduction in sea ice means lesser albedo from the Arctic oceans, which will therefore absorb more heat, as well as higher evaporation and greater heat exchange between ocean and atmosphere. The book *A Farewell to Ice*, by English oceanographer Peter Wadhams, provides a complete picture of the role played by ice and Arctic marine circulation in the context of global warming.

The reduced sea ice cover has important consequences for commercial routes in the Arctic. The famous passages to the north-west along the Canadian coasts, and the north-east along the Norwegian and Siberian coasts, are becoming possible lines of communication between Asia, Europe and America, with the associated not indifferent issues of national sovereignty and contention over the extension of territorial waters. The 'Arctic Silk Road', as it is sometimes called, links China and Europe through the Arctic seas, while in Iceland an enormous commercial port is currently under construction for ships that will cross the Arctic as part of the Finnafjord project.

So, we have a world that is opening up, not just because of naval routes but also because of the resources that have become easier to exploit (such as the hydrocarbon deposits). In recent years, the political tensions among nations interested in the Arctic have grown, as highlighted by the various declarations made by certain governments.

What's more, the reduction in ice cover and the extension of the ice-free season are changing the entire Arctic ecosystem, from the characteristics of algal blooms to the cod populations, with consequences for the local economy and fishing.

The image of a polar bear (*Ursus maritimus*) balancing on an ever-reducing block of ice is an iconic one. Though it is a little too dramatic, it is symbolic of a real situation. Polar bears use the platforms of floating ice to hunt seals and to move around, and the loss of sea ice seriously complicates their lives. Counting polar bears is a complex task and their numbers are difficult to estimate. Generally, these bears form distinct subpopulations, and the majority of those for which we have sufficient data are in decline. Currently, the International Union for Conservation of Nature (IUCN) reports a global population of between 20,000 and 25,000 polar bears and the species is considered 'vulnerable', a term that indicates an intermediate level of risk.

An important effect of the reduction in sea ice is that the polar bears increasingly tend to stay on land. They are splendid and mighty animals, but they are also extremely dangerous and it is increasingly common to meet one whilst working in the field. We were forced more than once to beat a rapid retreat due to the arrival of a bear when we were taking measurements in the Bayelva basin, and, recently, a bear attacked the camp in front of Longyearbyen airport on Svalbard. As they spend more time on land, the bears have developed the habit of preying on nesting marine bird colonies, causing their reproductive success to diminish. Though the bears aren't at immediate risk of extinction, the forced changes to their habits could instead threaten the survival of other species. As we know, ecosystems are formed by a network of interconnections in which just one change, no matter how innocuous it might seem, can propagate in often unexpected ways and affect the entire system.

Sinking into the Mud

The permafrost, which we have already encountered, is the perennially frozen soil at the poles and high in the mountains. It can be found in more than 20 per cent of the land in the Northern hemisphere and can reach thicknesses of up to 1 kilometre, as is the case in northern Siberia. The geothermal heat coming from the Earth's interior stops it forming at any greater depth.

Above the perennial permafrost is a layer referred to as 'active', which has a thickness that varies between a few centimetres and a few metres depending on local climate conditions. In the active layer, seasonal variations in temperature cause a partial summer thawing, which means the frozen soil is replaced by an often swampy surface area that can sustain the development of vegetation. In some areas, especially under deep lakes that do not freeze entirely, the superficial layer remains defrosted all year round and is called 'talik'. Generally, the dynamics of the Arctic ecosystems are strongly linked to seasonal variations in the surface layers of the permafrost and in the frost–defrost behaviour of the soil.

The perennial permafrost is extremely resistant and can take the weight of stable infrastructure such as roads and habitations, acting as an anchor for the foundations of houses. It is almost like a secure rock on which you can carry out your daily activities. However, it is a rock that is liquefying. The temperature of the Arctic permafrost is rising and, as a result, the defrosted layer is slowly getting thicker. In many areas of the Arctic, regular monitoring is carried out in order to provide a clear picture of just how the situation is evolving. For example, the research group led by German hydrologist and geochemist Julia Boike has been providing an ongoing recording of the permafrost characteristics and the meteorological conditions in a research station in the Bayelva basin since 1998. Similar measurements are available throughout the Arctic, often in the framework of international collaborations and programmes such as T-MOSAiC of the International Arctic Science Committee (IASC); see www.t-mosaic.com.

The measurements carried out by several research groups have been analysed in a work that summarizes the findings, coordinated by German polar scientist Boris Biskaborn and in which researchers from all over the world have participated. In the polar regions with continuous and

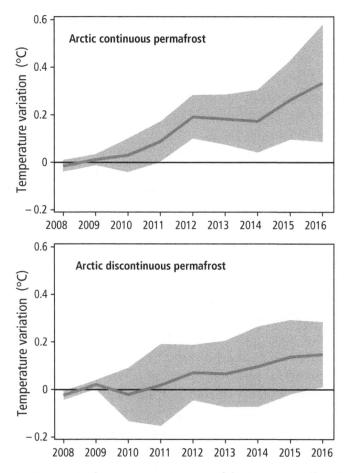

Figure 16.3 Increase in the average temperature of the Arctic permafrost in regions where it is continuous (above) and those where it is discontinuous (below). The shaded area indicates the variability of the measurements between the different sites considered. Taken from Boris K. Biskaborn et al. (2019), www.nature.com/articles/s41467-018-08240-4.

well-developed permafrost, the data indicates an average global rise in temperature of more than 0.3 °C in the period from 2007 to 2016 alone, around the depth where seasonal variation stops being visible. Someone, sadly, has said that we should be calling it perma-defrost instead.

The increase of 0.3 °C in a decade is, obviously, a particularly high one. In parallel with this rise in temperature, the superficial defrosted layer tends to become deeper, undermining the stability of the terrain and leading to the collapse of infrastructure, such as roads or habitations, which sometimes sink into the mud. Movement by land is made increasingly difficult and the main airport in Greenland will have to close due to the damage caused by the defrosting soil. On the ocean shore, the weakening of the permafrost makes erosion more common, with entire portions of coastline collapsing.

Moving from the land to the sea, the permafrost often continues beneath the seabed. It is a difficult presence to measure or estimate as it can be hidden by a thick layer of sediment. But, both at sea and on land, the most concerning characteristic of the permafrost is its very high content of organic matter and, therefore, carbon. Currently, as US ecologist Ted Schuur reported, it is estimated that around 1,000 billion tonnes of carbon are contained in the top 3 metres of the Arctic areas covered in permafrost. As a comparison, the quantity of carbon in the top 3 metres of soil in the entire rest of the world is estimated to be around 2,000 billion tonnes.

Italian polar oceanographer at the Italian National Research Council (CNR) Tommaso Tesi, and his team, have demonstrated that, during the last deglaciation, permafrost thawing released enormous quantities of carbon dioxide and methane, contributing to the rapid rise in temperature between 14,000 and 7,000 years ago. The mechanism is simple: if the temperature of the permafrost rises, the bacteria begin their efficient act of decomposing organic matter with the subsequent release of methane and carbon dioxide, and Bob's your uncle!

The concern is clearly that this process might be triggered in the current period. The images showing bursts of methane from holes in the frozen surface of Arctic lakes, or the circular craters in the Siberian tundra, probably generated by the release of large methane bubbles from the permafrost below, tell us that this is not a remote possibility. It really would be a serious problem, because the widespread decomposition of a good chunk of the organic matter that is currently frozen would lead to an even faster rise in global temperatures. Furthermore, there is the possibility that polluting substances, such as mercury, will be released from the frozen soil and that there could be a potential reactivation of pathogenic agents, viruses and bacteria currently trapped in the permafrost.

A Question of Mismatch

These great changes in the Arctic environment also have significant effects on the functioning of the polar ecosystems. The Arctic Council, which brings together the eight states with territory in the Arctic and various other nations acting as observers, has created a working group dedicated to the conservation of Arctic flora and fauna (CAFF), which carries out research and coordination activities in order to provide a regularly updated picture of the situation. Similarly, the Terrestrial Working Group of the IASC aims to improve communication and collaboration between researchers working on ecosystems and hydrology in the Arctic.

One initial aspect has to do with the northward movement of animal and plant species following the rise in temperatures. Changes in species distributions as a result of global warming are well recognized the world over, as has been widely illustrated in recent years by US biologist Camille Parmesan (now working in the UK) and her group.

In the Arctic, we can see how the shrubs and trees of the northern forests (Taiga) are colonizing areas that lie ever farther north, those areas previously dominated by the tundra. Similarly, vegetation pests are reaching ever higher latitudes. The European fox is now present in areas where previously there was only the Arctic fox. There are many examples of such movement, which is a part of the species replacement that occurs when environmental conditions change. An excellent review article on this subject was published in *Science* in 2009, written by US biologist and ecologist Eric Post and his collaborators. This work continues to provide a great deal of information on changes currently under way in Arctic ecosystems.

The point, however, is that a change that occurs too rapidly can trigger environmental instability, linked to a mismatch between the different components of the ecosystem. Due to rising temperatures, even the phenology of many plant species is changing. Arctic grass is flowering earlier, by weeks in some cases. If the other components, such as pollinating insects and herbivores, are not able to react with equal speed, a phase shift is created that is potentially damaging to the correct functioning of the ecosystem. In 2010, I worked with biologists at the University of Milan, Italy, on a research project looking into the dynamics of migratory bird species in the subarctic. Analysis of the data demonstrated how not all migratory species were able to adapt to the early arrival of spring-like conditions in Northern Europe associated with the increase in temperature. And it is precisely these species that are less able to adapt, or that show a greater mismatch with the changed climate situation, whose numbers are declining the fastest.

Generally, changes in environmental conditions can also influence the biodiversity of Arctic regions, amplifying other direct pressures such as pollution and the arrival of invasive species. The Arctic's endemic species and the ecosystems of the tundra find themselves today under significant stress. The CAFF has drafted a report on Arctic biodiversity that provides detailed information on how animal and plant life is changing north of the Arctic circle.

Adventures in the Tundra

Today, the Arctic is 'under special surveillance' precisely because of the large-scale changes taking place at top speed. The most powerful countries in the world have research stations in Alaska, Canada, Siberia, in the extreme north of Europe, in Greenland and on Svalbard. Oceanographic ships and ice-breakers sail the Arctic seas to analyse environmental changes, collecting data to understand what is happening. The little village of Ny-Ålesund, which faces the bay of Kongsfjorden on the island of Spitzbergen, has been transformed from a mining outpost to an international research centre hosting permanent stations belonging to many countries from all over the world, with cutting-edge instruments and laboratories. These include the Zeppelin Observatory of the Norwegian Polar Institute, dedicated to atmospheric measurements; the Gruvebadet laboratory dedicated to measuring the composition of the atmosphere, the characteristics of atmospheric particulate and the properties of snow; and the Amundsen–Nobile Climate Change Tower (figure 16.4), 34 metres high and capable of measuring temperatures, winds, the radiation balance, the composition of the atmosphere, the flow of carbon and water vapour and many other meteorological and climate parameters.

The Arctic tundra is a particularly important object worthy of careful study to understand how it is reacting to climate change. The growing depth of the active layer above the permafrost, the rise in temperatures,

Figure 16.4 The Amundsen–Nobile Climate Change Tower (CCT) installed by the Italian CNR close to Ny-Ålesund on the island of Spitzbergen. The CCT is 34 metres high and it measures meteorological and climate parameters and carbon, water and energy fluxes between the surface and the atmosphere. Author's photo.

the retreat of the glaciers and the difference in groundwater circulation are all elements that can influence the dynamics of the tundra ecosystems. The regions left free of ice are quickly colonized by the biological crust and, later, by vegetation. In the areas that have been vegetated the longest, there is a tendency for vascular plants to expand at the expense of mosses and lichens, bringing with them the possibility of a different composition of photosynthetic communities. The work by a great number of researchers, coordinated by ecologist Isla Myers-Smith at the University of Edinburgh, analysed satellite data and field measurements to demonstrate a noteworthy complexity in the Arctic tundra's response to global warming, with an alternation between 'greening' and 'browning', and different behaviour depending on the area and across different spatial ranges.

These changes can have significant effects on the carbon flow between soil, vegetation and the atmosphere, determined by the balance between the absorption of carbon dioxide by photosynthesis and its emission through the respiration of plants and the soil, and the decomposition of organic matter. One issue that remains unresolved is understanding whether there is a possibility that the tundra ecosystem could become a further source of atmospheric carbon dioxide. The work of physicist Marta Magnani from Italy's CNR and her team has, for example, identified the main drivers for carbon exchanges in the tundra around Ny Ålesund and provided the basis for extending this description to other areas of the Arctic. A review article written in 2021 by a large number of polar scientists and coordinated by Norwegian ecologist Åshild Pedersen summarizes the results of more than 25 years spent studying the environment on Svalbard.

There is much more to say about Arctic ecosystems, but we must remember one thing: the terrifying rise in fires at high northern latitudes. As we saw in chapter 15, in 2019, 2020 and 2021, enormous expanses of forest in British Columbia, Canada, were burning in a region that is traditionally untouched by fires. Many Siberian fires have occurred on land containing large quantities of peat, which is rich in organic matter and has continued to burn for extended periods of time. Facilitated by rising temperatures, the fires emit high concentrations of carbon dioxide, amplifying the greenhouse effect and releasing enormous amounts of soot into the atmosphere, which, when deposited on the snow, contributes to a decrease in the albedo and favours heat absorption. Arctic fires are a product of global warming that is rightly given a great deal of attention today, attention that aims to reduce its extent and impact on the environment.

Restless Vortices and Lazy Currents

As we commented at the beginning of the chapter, one reason for interest in the high latitudes is that the events seen in the Arctic are not confined to the polar regions.

In the troposphere at mid and high latitudes, around between 5 and 9 kilometres above sea level and between 40 and 50 degrees latitude north, the circulation system known as the circumpolar vortex develops. This is an area of strong wind at altitude that is associated with the atmospheric jet stream (which we have already discussed in chapter 7) and which intensifies in the winter and becomes weaker in the summer. On top of this, in the stratosphere, during the winter, another polar vortex is formed that is more localized above the Arctic, and that has a velocity that grows with altitude until it reaches around 50 kilometres above sea level. The vortices have little to do with one another and both have intense winds that blow from west to east. The tropospheric vortex is more irregular and characterized by huge planetary waves that deform its shape. Figure 16.5 shows the schematics of the two polar vortices in the Northern hemisphere. Similar vortices exist in the atmosphere of the Southern hemisphere, where the stratospheric vortex is one of the protagonists in the development of the hole in the ozone layer that developed a few decades ago.

The tropospheric vortex separates the cold air at the pole, which is generally a region of low pressure, from the warmer air at mid and low latitudes. When a particularly intense wave deforms the vortex, in some regions the polar air descends to mid-latitudes, whilst in others, warmer air moves towards the subarctic regions. Part of the winter variability of

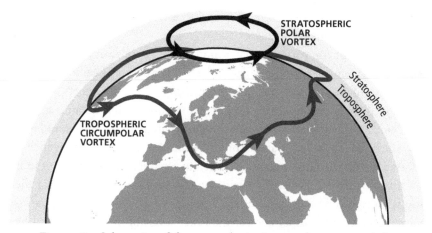

Figure 16.5 Schematics of the tropospheric circumpolar vortex and the stratospheric polar vortex, which is more regular and localized. Taken from the website of the Waugh Research Group, https://sites.krieger.jhu.edu/waugh/research/polarvortex.

the weather at mid-latitudes is due precisely to these fluctuations in the vortex.

Over recent decades, the circumpolar vortex has become more unsettled, with more marked waves. In this context, the irruptions of polar air to the south and warmer air in the poles become more frequent and more intense. On the same day, there can be a deep freeze in North America and spring-like conditions in Northern Europe, as happened in the winter of 2013–14, when New York was shut down by a snow storm while newspapers reported how, in Scandinavia, the bears were emerging from hibernation due to the high temperatures in the middle of winter. Similar events happened again in more recent years, such as the Arctic blast and extreme cold experienced in large areas of North America in December 2022.

Variability is one of the key words when it comes to global warming. Not only does the latter bring higher temperatures, but much more unstable weather conditions, which can make forecasts and planning for activities less certain. Presumably, the amplification of Arctic warming plays a central role in making the circumpolar vortex more unstable and therefore in causing a greater incidence of extreme events at mid-latitudes. However, we do not yet understand these mechanisms in detail. This is the subject of much research and, in 2019, the journal *Nature* published a collection of scientific articles looking at this exact subject from various perspectives.

If the atmospheric vortex becomes more lively, Arctic warming could, on the contrary, render marine circulation lazier. As we have already discussed, the release of fresh water through the melting of the ice sheets and the thawing of the permafrost can inhibit the sinking of the oceanic surface water and, ultimately, slow the thermohaline circulation. This would also slow the branch bringing waters from the south back to the north, which is associated with the Gulf Stream, and therefore reduce the transport of heat towards Northern Europe. That said, we must not forget that part of the heat is transported by the atmosphere and greater variability in the circumpolar vortex could compensate for the lower levels of heat transported by water, with a greater flow of heat and humidity by the winds, as suggested by the work of US climatologist Richard Seager and his team.

But just how present are the effects of the reduction of the Gulf Stream? In the first months of 2021, in her doctorate thesis at the

Potsdam Institute for Climate Impact Research (Potsdam-Institut für Klimafolgenforschung, PIK) in Germany, climatologist Levke Caesar (together with other researchers) re-analysed the available data through both direct measurement and past reconstruction, and concluded that, halfway through the twentieth century, Atlantic thermohaline circulation began to weaken, reaching its lowest point of the last 1,000 years. Perhaps this variation is part of the seesaw discussed by Wallace Broecker, though this time it is disturbed by a new player that is throwing as much carbon dioxide as it can up into the atmosphere. Research is still ongoing and new results become available every month. In any case, the Arctic is not a separate world, and what happens in the far-northern regions is rapidly reflected in the rest of the globe. Each year, the NOAA in the US publishes a report on the main scientific results on changes in the Arctic, which is useful for keeping us up to date on the research under way.

17

The Mountain Heat

He walked slowly along the trail after years of being away, expecting to find the glacier of his youth behind every curve. And yet, he did not. Where, as a child, he had practised climbing the icy walls that were more than 10 metres high with crampons and an ice axe, now there were only lakes and expanses of mountain flowers. The glacier could be glimpsed in the distance, reduced to a minuscule mass rendered grey by the detritus covering it. And yet only 40 years had passed, not centuries.

Mountainous Lands

Mountains occupy around 20 per cent of the land, a fraction of our planet that is by no means negligible and that currently hosts more than a billion people. In many cases, the mountainous regions act as water towers for the flatlands, and are, therefore, far more important than their geographical extension might suggest. As with the Arctic, with which they have a number of similarities, 'what happens in the mountains doesn't stay in the mountains', but, rather, has an impact on all of the surrounding areas.

The data and observations available indicate that mountainous areas are particularly sensitive to the effects of global warming, to the extent that they are considered one of the hotspots of climate change. In 2003, Swiss climatologist Martin Beniston published a wide-reaching summary of the impacts of climate change in the mountains. In 2008, the UN General Assembly highlighted that 'mountains provide indications of global climate change through phenomena such as modifications of biological diversity, the retreat of mountain glaciers and changes in seasonal runoff, that may impact major sources of freshwater in the world', recognizing that 'sustainable mountain development is a key component in achieving the Millennium Development Goals in many regions of the world'. More recently, the IPCC document 'Special Report on the Ocean and Cryosphere in a Changing Climate (SROCC)', published in 2019, includes a chapter that provides an exhaustive and up-to-date picture of what is happening at altitude all over the globe.

One reason for the mountains' particular sensitivity to increases in temperature is the role played by the zero degree isotherm: if, in some areas of the planet, the average temperature rises from -20 to -18 °C, or from 20 to 22 °C, this warming will have significant effects locally but probably less immediately than a passage from -1 to +1 °C. In this case, the ice present before the warming will melt and the environment will change completely. This leads to the first significant effect of the rise in temperature: the zero degree isotherm is shifted to ever higher altitudes, and the entire system reacts to this change. In addition, many organisms that live in the mountains are extremely specialized in managing the difficult conditions at altitude and struggle to adapt to the increase in temperature. Changes to the environment, the water, the ecosystems and biodiversity all contribute to a rapid modification of the world of mountains.

Amplified Warming

As in the Arctic, the warming in the mountains seems to be happening faster than in the surrounding areas, even if the observational evidence is more fragmented than in the polar regions. In the Alps, for example, the rate of warming has been around 0.5 °C every 10 years, as indicated by the European Environment Agency (EEA). In many mountain areas, we have witnessed what has been called Elevation Dependent Warming (EDW), a rise in temperature that gradually becomes more intense at higher altitudes and becomes particularly significant around the zero degree isotherm. This phenomenon, particularly evident in European and Asian mountains, has been proven by many researchers, with US climatologists Henry Diaz and Raymond Bradley being some of the first to highlight it in 1997. It has also been discussed in detail by a working group of the Mountain Research Initiative (www.mountainresearchinitiative. org), coordinated by UK climatologist Nicholas Pepin. The data from the research stations on land and the satellite observations is less homogeneous than the information available for the Arctic, but the picture that emerges is still fairly clear.

Climate models, which we will discuss in more depth in the next chapter, all detect a warming that increases with altitude, as discussed by Italian climatologist Elisa Palazzi and her team in a high-spatial-resolution study for the Alps, the Himalayas – Tibetan Plateau region, and the Rocky Mountains in Colorado, USA. The study was then extended by Brazilian climatologist Osmar Toledo and his team to the tropical and subtropical Andes. Previous modelling work has analysed the situation in larger areas of the Rocky Mountains, the mountainous regions of Korea, in Mexico and in Central America. Substantially, despite significant geographical differences, the higher the altitude, the greater the warming effect seems to be.

The mechanisms responsible for this behaviour are varied and not yet entirely understood, and they can vary according to the season and the geographical area. An important role is undoubtedly played by changes in the albedo, with the ice-albedo or snow-albedo feedback mechanism we already saw in the Snowball episodes and the triggering of the glaciations over the last 3 million years, as well as in the current Arctic amplification. In the case of Elevation Dependent Warming, the effects are more limited but equally significant. The increase in the altitude of the zero degree isotherm leads to the melting of large quantities of ice and snow in areas that were previously glacialized. This melting exposes the soil and rocks below, which are darker and absorb more solar heat. These warm up, amplifying the rise in temperature and the deed is done.

But the change in albedo is not the only factor at work. Another element is linked to the rise in infrared radiation that returns from the atmosphere to Earth, essentially a localized increase of the greenhouse effect. The main cause seems to be an increase in atmospheric water vapour, another positive feedback that we have already encountered. This mechanism is more complex but, to put it simply, we can say that the atmosphere in the mountains warms up and becomes more humid, holding more water vapour, which in turn amplifies (locally in this case) the greenhouse effect. This mechanism has been discussed by US climatologist Imtiaz Rangwala with regard to the Tibetan Plateau, and has been identified in many other mountainous regions. Similarly, the rise in atmospheric aerosols and, in particular, black carbon (soot) can locally modify the properties of the atmosphere and induce a warming effect. Furthermore, when the black carbon is deposited on a snowy surface, it absorbs the solar heat and causes the snow to melt further. For anyone wanting to know more about Elevation Dependent Warming and its causes, in 2022 Nicholas Pepin and his co-workers published a comprehensive work that looks at mountainous regions all over the world.

Ice that Flows ...

Mountain glaciers are large, slow rivers of ice that flow from high altitudes towards the valleys below. A glacier's size is generally controlled by two factors: the accumulation of ice, associated with snowfall, which occurs in winter for glaciers at mid-latitudes, and ablation in the warm season, which causes the ice to melt and is controlled by temperature and solar radiation. If accumulation wins, the glacier grows, whereas if ablation prevails, the glacier retreats. The mass balance of a glacier measures precisely this difference between accumulation and ablation. A positive balance means the glacier is growing; a negative balance indicates its shrinkage.

For fairly extensive glaciers, accumulation happens for the most part at higher altitudes, where the temperature is lower, while ablation happens farther down, close to the front of the glacier. The Equilibrium Line Altitude (ELA) indicates the altitude at which the two processes (accumulation and ablation) cancel one another out. Above this line, the ice accumulates, while below the ELA the ice tends to melt. This difference means that the glacier 'swells' at higher altitudes, where the snow accumulates and is slowly compacted by the new layers, gradually turning into ice. The older ice, sometimes referred to as 'blue ice', is characterized by the almost total absence of trapped air, has larger ice crystals and is particularly compact. Thanks to the laws of gravity, the ice that accumulates at higher altitudes begins a slow descent, a bit like a very viscous fluid. As it flows towards the valley, the ice moves below the ELA and this is where melting begins. The ice becomes water, which then flows like a proglacial stream towards the valley floor. So, while a glacier might seem static, it is actually engaged in a continuous downward movement.

The situation is, of course, more complex than what we have described. For example, in many mountain glaciers there is a thin layer of water underneath that makes its downward flow much faster. Other times, the glacier flows on a layer of fine sediment that can be easily deformed and, once again, favours the glacier's movement. Inside the glaciers, a network of cavities is created through which the meltwater flows, and in the largest glaciers spectacular tubular caves are formed, bathed in a blue light. Splendid, yes, but extremely dangerous to explore due to the constant instability caused by the glacier's movement. Some glaciers show signs of an oscillatory instability, meaning that, every so often, the ice that accumulates at the top rapidly slides down towards the valley, causing a sudden lengthening of the glacier that is followed by a period of accelerated melting of the ice that has reached the lower altitudes. These are known as glacial surges, a fascinating phenomenon that occurs in many mountain ranges the world over, sometimes with disastrous consequences.

In general, if the climate is constant, an equilibrium is reached between accumulation and ablation, meaning the glacier does not tend to change its extension (length, thickness and width). New ice is formed each year higher up, which then flows towards the valley and melts below the ELA. Due to climate changes, however, the ELA can rise or fall. If the equilibrium line drops in altitude, then accumulation will be greater than ablation and the glacial volume increases, while, if the ELA rises, then ablation prevails, the mass balance will be negative and the glacier will tend to melt much more than it grows during the season with abundant snowfall. In many countries with glacialized areas, there are initiatives monitoring the cryosphere and, on a global scale, the World Glacier Inventory (WGI, nsidc.org/data/glacier_inventory) keeps a quantitative track of how glaciers are changing around the world.

... And that Melts

Due to global warming, which, as we have seen, is amplified in mountainous regions, over recent decades the mass balances of glaciers have become significantly negative almost everywhere. Figure 17.1, taken from the SROCC report by the IPCC (already cited), shows the average progress of mass balances measured in eleven mountainous regions from all over the globe, confirming the continuous loss of mountain ice.

In 2018, climatologist Martin Beniston, whom we have already mentioned, and a vast group of co-authors published an extended

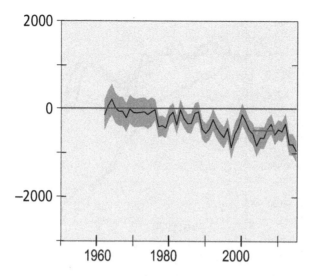

Figure 17.1 The average progress of mass balances in eleven mountainous regions all over the globe, confirming the continuous loss of mountain ice, taken from the SROCC report by the IPCC, based on results reported by Zemp et al. (2019). The vertical axis is in kg/m^2 per year. The grey band indicates the range of variability and the short horizontal line indicates the estimates made by Gardner et al. (2013).

article in which they discuss the responses of the Alpine cryosphere to global warming. Figure 17.2 shows, for example, the variation in length of several large alpine glaciers compared to their extension when monitoring began. It is clear once again that there is a widespread retreat, associated with a predominance of summer melting over winter accumulation, and a net loss of ice.

One unique case is those glaciers covered in debris, which, if thick enough, can slow the melting, insulating the ice from the surrounding environment. Many glaciers in Karakoram are like this, and we will return to the anomalous behaviour of this mountain range shortly.

On the southern slopes of the Alps, the Miage glacier in Val d'Aosta, shown in figure 17.3, is a European example of a 'black' glacier. The length of the Miage has not changed over recent decades. However, the total volume of ice has diminished due to a negative mass balance and the ice flow seems to have slowed, as discussed by English glaciologist Anne Stefaniak and her team in a 2021 article. Indeed, for many glaciers,

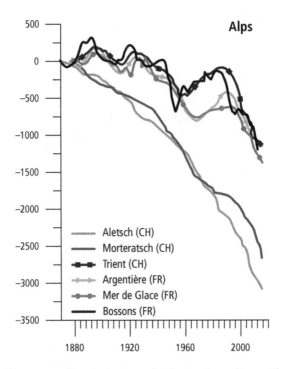

Figure 17.2 Variation in length (in metres) of a number of large Alpine glaciers, compared to the length measured when observations began. Taken from Martin Beniston et al. (2018).

which tend to be smaller than the Miage, the final phase of their existence includes a long period in which the ice is covered in rock fragments and becomes inert, with no further obvious flow. An ancient vestige from a different time. But, for now, the Miage is still alive.

Water Towers

The retreat of the glaciers has several repercussions on the availability of water resources, particularly in the drier months when the melting ice feeds the rivers and streams in the lean period. Equally important is the reduction in snow cover and the early melting of seasonal snow. In recent decades, the average depth of winter snow cover has decreased in almost all regions of the Alps, not so much due to a reduction in winter precipitation but because of higher temperatures, which encourage melting

Figure 17.3 An image of the Miage glacier covered in debris, in 2014. The icy tongue descends from Mont Blanc and curves to the left (to the right for the spectator), dividing itself into three main branches. Note the trees on the sides of the glacier and the small lakes that have formed on the covering of fragmented rock on top of the glacier. Author's photograph.

and increase the percentage of precipitation that falls as rain rather than snow. The early melting of snow sends a great deal of water back into the rivers in spring but increasingly less as the seasons progress, further reducing river flow in the hotter months and contributing to hydrological stress in the valley areas, which is amplified by the high summer temperatures. It is also possible to observe a reduction in snow cover in other mountainous areas, such as in the Hindu-Kush–Himalayan region and in the Andes during the dry season. In certain cases, such as in the Rocky Mountains in Colorado, the signs are more complex and lead to changes in the monthly distribution of winter snow precipitation.

The main reason we are interested in how the water cycle is changing in the mountains is that, in many cases, mountainous regions act as water towers for surrounding areas, which depend on them for the water used in agriculture, industry and for human consumption. Dutch hydro-meteorologist Walter Immerzeel, Swiss hydrologist Daniel Viviroli and their team have explored in great depth the concept of mountains as water towers, analysing the importance of mountainous areas in the different parts of the world. Their 2020 article summarizes what we know about this subject, discussing also the vulnerability of various regions to climate change. From this analysis, we see that the water coming from the mountains today affects, directly or indirectly, almost 2 billion people, and that the most important areas – from the perspective of their capacity to provide water – are also the most vulnerable. In these areas, changes to the hydrological cycle, the early melting of snow and the loss of glaciers can bring water stress, social and geopolitical instability and a general difficulty in water management.

The area Hindu Kush – Karakoram – Himalaya – Tibet clearly emerges as among the most important, as it is the source of the main rivers in Asia (from Tibet, for example, flow the Indus, the Yellow River, the Yangtse and the Mekong) and is surrounded by extremely densely populated foothills characterized by significant geopolitical tensions. The reduced availability of water, the risk of a rise in landslides due to rainfall substituting snow, and ever shrinking glaciers cause serious environmental and social problems that are often difficult to resolve, but which the local populations are attempting to tackle. For example, in Ladakh in the north-east of India, the dry season is winter and spring, before the monsoon arrives. During the autumn and the beginning of the winter, the village's inhabitants collect the water in various kinds of tanks, where it then freezes before melting the next spring, releasing the water needed for agricultural practices. Today, it is typical in this region to find ice stupas, ice structures similar to small Buddhist temples that can preserve the precious water until the spring months.

Water, water, water. This is the big issue when it comes to climate change – and not only that. It is necessary for life, agriculture, for society. It is the reason for wars and peace treaties, for invasions and negotiations. Throughout the world, not only in the mountains, the issue of water occupies a central position in strategies for adapting to climate change and sustainable development. We will return to this, but the availability of water (too much or too little) and its qualities already are, and will increasingly be, among the critical issues we will be forced to face in the coming decades.

Erratic Monsoons

In many areas at mid-latitudes, the autumn and winter are cold and characterized by abundant precipitation, whilst summers are hot and dry. In these conditions, the glaciers are recharged by winter snow accumulation and melt through ablation during the summer months. But from what we have seen in Ladakh, we realise that in the Himalayas the situation is different: the hot summer months, dominated by monsoons, bring both snow at high altitudes and melting towards the valley. So, for the Himalayan glaciers, accumulation and ablation happen simultaneously at different altitudes.

In Himalaya, in Tibet and in a large part of south-east Asia, the atmospheric dynamics are dominated by monsoons. The Indian monsoon system is probably the largest and most famous in the world, and many aspects of the region's society and economy depend on it. Around three-quarters of the total annual precipitation falls during the monsoon season between June and September, providing the water resources necessary for irrigation, the production of electric energy and other human uses.

As with other monsoon systems, the dynamics of the Indian monsoon are linked to different levels of heating of the land and sea in different seasons. In the summer, the continental masses, and in particular the whole of the Tibetan Plateau north of the Himalayan range and the Thar desert on the border between India and Pakistan, warm much more than the surrounding Indian Ocean, also heating the air that is in contact with the surface. This generates intense convective activity in the atmosphere that lifts the air upwards. As a result, the humid air masses above the Indian Ocean are pulled towards the continents, which are characterized by lower atmospheric pressure than that above the ocean. The Himalayan range then acts as an obstacle for these south-west winds, filled with humidity because they have crossed the Bay of Bengal, causing heavy rain every summer on the most eastern side of the mountain ranges and a large part of south-east Asia. Given the importance of this subject, there are numerous scientific studies looking at the Indian monsoon. Here, I will cite only one article, published in 2017 by Chinese climatologist Pin Xian Wang and his co-authors, who were participating in a specific working group of the PAGES initiative mentioned earlier. This work discusses in detail the properties of the south-east Asian monsoon and its variability over time.

The issue of possible changes to the Indian monsoon as a result of global warming and other environmental modifications is clearly very important, not least due to the enormous number of people who will suffer the consequences. In recent years, the monsoon seems to have become stronger and more irregular. In 2021, a group of Chinese and US climatologists, led by Qinjian Jin from the University of Kansas and his team, discussed how some changes to the monsoon are due to the rise in aerosols, which has been particularly intense in this part of the world. Aerosols change the optical properties of the atmosphere, modifying the processes of absorption and re-emission of solar radiation, leading to possible changes in atmospheric circulation. Analysing the results of the latest generation of climate models, climatologist Anja Katzenberger and her team at the Potsdam Institute for Climate Impact Research confirmed the Indian monsoon's growing tendency to become stronger and more erratic. This concerns not only changes to its mean characteristics, therefore, but (here as in many other cases) an increased variability with all the planning difficulties this brings for the management of water resources.

Winds from the West

To the west of the Himalayas, the Karakoram range is home to some legendary peaks, such as K2. And it is precisely in Karakoram that we find an unexpected exception to the widespread retreat of glaciers. In this mountain range, the balance between accumulation and melting is slightly positive and glaciers seem to be in good health, generating what is often referred to as the Karakoram anomaly.

As ever, the causes are many. One is that the Karakoram glaciers are covered in rocky debris that can help protect the ice from melting. The stability of these glaciers has been attributed to the 'insulating' effect of the debris layer, which makes the 'black' glaciers subject to different ablation dynamics than the typical 'white' glaciers, as discussed by German climatologist Dirk Scherler and his team in 2011. But this is unlikely to be the main reason. In the Karakoram region, the summer monsoon circulation is joined by the winter circulation linked to perturbations at mid-latitudes that come from the west – the Atlantic and Mediterranean, in particular. Along their route, these perturbations collect humidity over the Caspian and Arabian seas, bringing precipitation to Karakoram typically between December and April, as discussed in detail in 2014 by Italian researcher Luca Filippi in his doctoral thesis.

Winter precipitation associated with perturbations coming from the west, known as Western Weather Patterns, are the main water source for the Karakoram glaciers. These glaciers are, therefore, similar to the Alpine ones, accumulating snow in the winter and melting in the summer. And yet the measurements provided by the climate stations indicate that, presumably following changes in atmospheric circulation, in the region of Karakoram winter precipitation has risen and summer temperatures have slightly fallen, two factors that can explain the weakly positive values of the mass balance observed for these glaciers. An extremely interesting puzzle, whose solution can help us better understand the glaciers' response to climate changes, and which in 2020 was summarized by Swiss glaciologist Daniel Farinotti and his colleagues in an article on the subject.

Life at Altitude

Due to the heterogeneous nature of their physical, topographical and climate characteristics, mountainous regions create a mosaic of habitats that vary with altitude and generate an incredibly high level of biodiversity. At the same time, mountains are home to some of the rarest ecosystems in the world, capable of withstanding extreme conditions. The plant and animal populations that live at altitude are often small and isolated, made up of species that have adapted to the low temperatures and that, given their limited capacity for dispersion, easily fall prey to local extinction. For all of these reasons, mountain ecosystems are particularly sensitive to changes in the climate and physical environment.

One effect is that the rise in temperature leads the mountain species to move towards ever higher altitudes, the mountain version of the poleward displacement described, for example, in the works of Camille Parmesan, already mentioned in chapter 16. Though the evidence is still fragmentary, there are indications of an upward altitudinal shift of alpine plants in many mountain areas, from the Swiss Alps to the Sichuan in China, the Green Mountains in Vermont and many more. Similarly, the tree line is rising steadily in almost all the mountainous areas of the world, but in this case it is difficult to distinguish the effect of warming from the probably more important phenomenon of the abandonment of high-altitude pastures and changes in land use and occupation, with the consequent re-colonization of the pastures by trees and shrubs.

Many animals, too – especially invertebrates, but also birds and mammals that in recent decades were confined to lower altitudes – are now present at ever higher elevations in the Alps, Central America, the Rocky Mountains and the tropical Andes. This is happening, for example, with the pine processionary moth, a species that is highly destructive to plants which is colonizing relentlessly at higher altitudes. The problem, however, is that, as elevation rises, the available space decreases, and beyond a certain altitude there is simply no more mountain. At times, it is sadly suggested that these species are getting on an 'escalator to extinction'. In 2018, Canadian biologist Benjamin Freeman and colleagues described an example of this kind for a number of bird species in the tropical Andes. In the Australian Alps also, where the highest peaks reach around 2,000 metres, mountain vegetation seems to be struggling to adapt to the new, warmer conditions, and it cannot move any higher because there is nowhere higher up to go, as demonstrated by plant ecologist Meena Sritharan in 2021 during her research doctorate at the Australian National University in Canberra.

The mixing with species who have come from lower down also directly threatens the survival of less mobile organisms or those with specific environmental requirements, bringing about a potential loss of biodiversity. The research group led by biologist Ramona Viterbi in Italy's Gran Paradiso National Park (with which I work) has been carrying out quantitative monitoring of different kinds of invertebrates for 15 years now, in order to estimate changes in Alpine biodiversity following the rise in temperatures and variations in precipitation. The results essentially show a greater risk of losing precisely those endemic species that are under threat, more typical of these environments, in favour of generalist and more ubiquitous species.

Ibex and Prairies

Precisely in the Gran Paradiso National Park, which celebrated its one-hundredth anniversary in 2022 and which was Italy's first national park, we find the original nucleus of alpine ibex (*Capra ibex*), the park's symbol, from which hail all the individuals that have repopulated the Alps over the years (see figure 17.4). Let's look at what determines the dynamics of this population, as discussed in 2004 by US researcher

Figure 17.4 An adult male alpine ibex (*Capra ibex*) at the Gran Paradiso National Park, Italy. Author's photo.

Andy Jacobson – today at the Cooperative Institute for Research in Environmental Sciences (CIRES) at Colorado University in Boulder – together with myself and the Park's biologists.

Our starting point is the availability of a long series of annual counts of the Park's ibex population, which began in 1956 and continues today. Data of this kind is essential when analysing the causes and mechanisms of ecosystem change and to estimate its possible future evolution, but it requires great dedication by the park rangers and staff, and those of us analysing the data must recognize their hard work.

From this data and the concurrent weather and climate measurements, we can see that the ibex population in Gran Paradiso (between around 3,000 and 5,000 individuals) is principally controlled by the combination of the size of the population and average depth of winter snow. The 'bottleneck' is winter, because this is the period when the ibex have to feed by digging in the snow or finding the few rocky areas exposed to the sun. So, if there is too much snow and the population is too big, the challenge of finding food becomes extreme and many adults do not make it. Vice versa, years with little snow favour population growth. In the mid-1980s, when the Alpine snow cover began to reduce drastically, the ibex population of Gran Paradiso grew and reached its maximum of around 5,000 counted individuals.

But then, halfway through the 1990s, it collapsed, reaching levels that were below those measured before the increase, due to the fact that the newborn ibex struggled to survive their first winter. The average percentage of surviving year-old ibex kids almost halved. There are various possible reasons for this debacle. The increased survival rate of the adults, linked to the reduction in snow, could have led to a population of older individuals that were less capable of producing offspring that were sufficiently robust. Or, instead, it might have been an early blossoming of the plants at altitude, which in the Alps, as in the Arctic, is becoming increasingly evident.

Then, in summer, when the mothers would need high-quality food in order to produce sufficiently energy-rich milk and the kids would start feeding on grasses, it is possible that the vegetation had already passed its best and was not nutritious enough. In this case, it is interesting to note the two-fold effect of the snow, which limits the survival of the adult ibex in the winter, but that, when it is scarce, can lead to imbalances in the complex network of relationships that governs an ecosystem.

This hypothesis was explored in detail by English zoologist Nathalie Pettorelli and her colleagues, who in 2007 analysed satellite data on the timescale of vegetation development in four different mountainous areas – three in Canada and one in Gran Paradiso – linking the changes in plant productivity and phenology with survival levels of the offspring of bighorn sheep (*Ovis canadensis*) at Ram Mountain and Sheep River, mountain goats (*Oreamnos americanus*) at Caw Ridge, and the ibex at Gran Paradiso. The three Canadian areas are all in Alberta, more or less in the region of the Rocky Mountains. Analysis of the data showed that the more rapid the growth of vegetation, the lower the survival rates of the offspring of bighorn sheep and ibex, and the slower the growth of the offspring of bighorn sheep and mountain goats. As the earlier growth of vegetation can imply a shorter high-quality foraging period, these results seem to confirm the possible mismatch between the development of vegetation caused by the warming of the mountainous regions and the response of wild herbivores.

Let us return, then, to the plants, soil and alpine prairies that sustain the ecosystems at high altitudes. The soil, rocks, organisms and groundwater form one large, single whole, the 'Critical Zone' we discussed in chapter 6 – the layer in which all the processes regulating terrestrial ecosystems take place. High in the mountains, the Critical Zone is characterized by the presence of high-altitude pastures and alpine tundra. In order to understand how these mountain ecosystems are changing, a few years ago our research group at the CNR installed, in collaboration with the Gran Paradiso Park, an observatory of the Critical Zone in the Nivolet plateau, measuring the fluxes of water and carbon dioxide, the chemistry of soil and water, and the composition of biological communities. In 2020, researcher Marta Magnani, during her research doctorate at the University of Turin, used this data to develop a numerical model to describe how the alpine Critical Zone is reacting to changes in the climate and estimating its capacity to sustain mountain organisms, including the ibex.

Measuring the Change

Changes in mountainous regions are multiple, rapid and complex. Different areas respond in different ways, and the great heterogeneity of geographical and climatic conditions can be seen in an often varied response, which makes it difficult to get an overall picture that works for all mountains. It therefore becomes essential to observe regularly and measure the changes over the years, with an effort that requires lots of time and great patience. But this is the only way that we can learn about what is happening.

For this reason, over recent decades various permanent observatories have been set up with the aim of monitoring and inspecting the changes to mountain ecosystems, in order to develop management methods capable of counteracting the adverse effects of global warming. Among these, it is worth mentioning the huge observatory of global changes in the Sierra Nevada in southern Spain, that in the Kalkalpen National Park in Austria, and the environment and ecosystem studies on La Palma in the Canary Islands. There is also the mountain observatory of ecosystems, hydrology and the Critical Zone at Boulder Creek, Colorado, and that of Catalina–Jemez in New Mexico and Arizona; the many Chinese observatories in Tibet, our planet's 'Third Pole'; as well as the Indian observatories in the Himalayas and many others, including the observatory created in Gran Paradiso National Park in Italy.

These observatories do not generally work in isolation, but belong to large international networks with the aim of harmonizing the measurements and information collected. So, let's not forget the International Long Term Ecological Research (ILTER) network, which is active all over the world; the National Ecological Observatory Network (NEON) in the US; the Chinese Ecosystem Research Network (CERN); and the Global Observation Research Initiative in Alpine Environments (GLORIA), the international network dedicated to permanent *in situ* measurements of the world's mountains, paying particular attention to the vegetation. We also have the international Critical Zone Exploration Network (CZEN), which comprises (though currently in a fairly informal way) many Critical Zone observatories in all areas of the planet. So many initiatives, so many networks – perhaps too many – that require further organization, and an attempt to harmonize all of these efforts to provide useful, coherent and complete knowledge on how the natural world is changing, information that can be used as a guide for models of the climate and its changes.

Models of the climate and forecasting the impact of climate changes – until this point, we have said very little about them, but that time has now come, bringing us closer to the end of our long journey through the history of the Earth's climate.

18

Digital Twins

The fragment of rock followed its orbit as the impetuous winds blew and the marine currents churned up its oceans. On that blue, ochre and green globe, the white clouds formed complicated arabesques, and ice sheets reflected the star's light. The biosphere extended throughout, modulating the cycles of the elements and the composition of the atmosphere. The climate of that planet was continually changing, due to modifications in external forcings, the vivacity of its own internal processes and the actions of its inhabitants.

In the silicon belly of the great computer, millions of processes carried out operations that gave life to the planet's digital twin, reproducing its most important characteristics. The data flowed continually from sensors and satellites observing the entire Earth and was assimilated in the computer codes. The digital twin tried to reproduce the climate in order to be able to predict its future behaviour, to face and mitigate negative changes and offer remedies for possible problems. But the young digital twin nevertheless remained nothing more than a metaphor. It dreamed of reaching the complexity and wealth of its sibling, older by 4.5 billion years, but its creators' knowledge, though vast, was not yet capable of spanning everything that existed in heaven and earth.

Thick Simulations, Tiny Similes

Up until this point, we have discussed the climate of the Earth based exclusively on data, provided both by direct observation and by paleoclimatic reconstruction, and on the laws of physics and biogeochemistry. But in order to understand the relationships between the various processes, and attempt to predict how they will behave in the future, it is necessary to transform our knowledge into a quantitative mathematical model that allows us to determine the state of the climate in response to the forces acting on the system.

The first discriminant factor involves the level of detail we want to achieve. In some cases, a simplified description is sufficient, including only certain aspects of our interest. We may want to study relationships of cause and effect, understand how a certain climate variable responds to the change in a forcing factor or another variable that acts on it. In this case, we can develop conceptual models that are not used to make direct predictions and which we may affectionately refer to as 'tiny similes', connecting things we hope are alike. The example of energy balance models (EBMs) based on the first principle of thermodynamics, as discussed in chapter 3, is a good example of this kind of approach. We can develop similar descriptions for the glacial–interglacial oscillations, the thermohaline circulation and the effect of vegetation on the climate, as happens with the Charney mechanism. We recognize this attitude also in the many conceptual approaches developed by meteorologist Edward Lorenz of MIT, who in 1963 was the first to observe chaotic behaviour in a simple deterministic numerical model, or by geoscientist Michael Ghil, whom we have already mentioned and who analysed many climate and paleoclimate processes.

For a more complete description of the climate system, however, the conceptual models are not sufficient. We must turn to more complex and complicated simulation codes that are capable of including many different components and which can reproduce a large number of aspects pertinent to our planet's climate. Here, we enter the realm of global climate models (GCM), which we will somewhat irreverently call 'thick simulations'.

Traditionally, global climate models come about from the union of various components: models for atmospheric dynamics and chemistry, the ocean, the terrestrial and marine cryosphere, the biosphere or the hydrological cycle. These individual components have been developed by different research groups hailing from a diverse range of disciplines and institutions. In the construction of a global climate model, great effort goes into making these components 'talk' to one another in order to build a coherent whole that can represent the Earth System in its entirety. Indeed, taking pieces from vastly different sources and gluing them together may seem a bit like a Young Frankenstein's game, but it can produce interesting effects ...

The tiny similes can provide an overall understanding of the processes at work, but there is more realism in the thick simulations. Figure 18.1

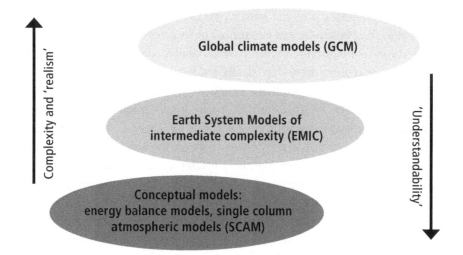

Figure 18.1 Schematic illustrating the hierarchy of climate models. This ranges from the conceptual models of specific processes, characterized by high understandability and low complexity and realism (the 'tiny similes' at the bottom on the left) to global climate models, which are complex and relatively realistic but more difficult to understand in detail. In the middle, we have Earth System models of intermediate complexity.

schematically depicts the hierarchy of climate models, also indicating an intermediary category: Earth System Models of Intermediate Complexity (EMIC). These include many aspects of the climate but are represented in a simplified way and, according to those who developed and use them, they are almost as comprehensible as the conceptual models and almost as realistic as the larger simulations. They are used predominantly for long-term paleoclimatic studies, or to verify the effect of the particular mechanisms in a more articulated environment than is possible with conceptual models, but their reliability is not always easy to gauge.

Models of the World

A global climate model is based on a collection of algorithms that resolve a large number of equations for the system's various components. As a first step, however, we must define which equations we intend to use.

For the atmosphere and the ocean, we use fluid dynamics equations (known as the Navier–Stokes equations) or some of their simplifications. With these two components, we have a detailed knowledge of the laws of physics (and chemistry) that ideally regulate their behaviour – even if there are other kinds of problems which we will look at shortly.

For many other components, the situation is much less satisfying. We do not know the determining equations for a forest's behaviour, and the details of water and energy flows through the soil or the dynamics of the hydrological network within a glacier are only partially known. Nor do we know how to accurately write the equations that describe the movement of wet sand. With all of these aspects, the basic knowledge available today (which uses 'first principles') is integrated by empirical formulas derived from data analysis and experiments carried out in the laboratory and in the field. In all of these cases, the equations we write are nothing more than a heuristic description of how the real system functions, and they improve the more we manage to capture the essential ingredients of the problem, eliminating the non-essential embellishments – in the hope, of course, that these do not reveal themselves to be crucial components for the task we have set.

The difference between tiny similes and thick simulations does not, therefore, lie simply in the approach but also in the number of processes we wish to describe simultaneously. Over the course of the years, from the first climate models that included only the atmosphere, moving slowly towards increasingly detailed descriptions with a greater number of components, global climate models have become more complex and reliable. The reports produced by the IPCC, which we have already discussed, allow us to explore the road travelled thus far by the science of climate models, from the versions created in the 1980s to the complicated simulations of the Earth System used today. The models are not and will never be perfect and there is still a long way to go, but it is reassuring to discover that the main results of the climate's response to the variation in driving forces, such as the growth in the concentration of carbon dioxide in the atmosphere, have substantially remained stable, becoming more precise with the development of newer generations of modelling descriptions.

The results of the various versions of global models are collected in the archives of the Coupled Model Intercomparison Project (CMIP) within the World Climate Research Programme. At the moment, fifth-gener-ation modelling products are available (CMIP5), while simulations for the sixth (CMIP6) are being completed. Of course, no lone researcher can single-handedly manage the extreme complexity of these algorithms and make the results available. Let's see, then, what carrying out a climate simulation actually means.

Equations on the Grid

Once the equations have been written and the way to describe the inter-actions between the various components has been ascertained (which in itself is not easy), we must then solve them. However, very few can be resolved analytically using paper and pen. We must therefore write approximation algorithms in order to deal with them in a numerical way with the help of a computer. At the moment, the calculation tools available are based on the logic of discretization, in the sense that they cannot deal with a space or time that is continuous but must be discre-tized on a spatial and temporal grid. In other words, in the climate represented by the model, we cannot describe what happens 1 centimetre away from us but must a priori define a spatial (and temporal) interval beneath which we cannot resolve the system's variability. The science of numerical simulation is, essentially, the development of increasingly efficient, accurate and reliable methods with which to translate an intrinsically continuous world into a grid description, without losing the essential aspects and, most importantly, preventing the discretized system from behaving too differently from the real one.

Let's see, then, how big a grid we need to describe the world. To simplify, let's consider only the atmosphere, which has a horizontal extension equal to the surface of the Earth and a vertical dimension, say, of 15 kilometres. The surface is worth $4\pi R^2$, in which R is the Earth's radius, which measures about 6,371 kilometres, so the result will be a little more than 510 million square kilometres. Multiplying this by the atmosphere's height, we obtain an atmospheric volume of around 7.7 billion cubic kilometres. If we wanted to have a spatial grid with a resolution of 1 metre, we must multiply the number we have just achieved by a billion (the number of cubic metres in a cubic kilometre). We will then have a grid with around 7.7 billion billion points. Suppose we have only one variable in each point of the grid, for example temperature, described in 'single precision' with 4 bytes (equal to 32 bits), we will have a total of around 30 billion billion bytes, or around 27 million terabytes, just in order to register a single global snapshot of the temperature in a fixed moment. And then there are all the other variables and all the other temporal moments. To be honest, this is not truly feasible with the instruments we have at our disposal today, and it will probably remain this way for a long while.

We must therefore reduce the grid's resolution. Today, global climate models only provide a detailed description for spatial scales larger than 20 to 25 kilometres for the most resolved simulations, and of 70–80 kilometres for most models. They provide no information on smaller scales. They do not, therefore, allow us to distinguish what is happening (or will happen) in my town as opposed to the next one over. The challenge here is clearly to make calculation algorithms increasingly efficient and speedy, designing new numerical solutions and developing computer architectures that are more efficient and adapted to dealing with this problem, in order to increase the resolution and reach ever smaller scales. But it will be difficult to get this down to a metre, and perhaps not even be necessary.

The Science and Art of Parameterization

The existence of a minimum scale beneath which a model cannot describe the system is, clearly, a serious problem. Not only because we completely lose the information on the dynamics of smaller scales, but also because there are many localized phenomena – such as storms, turbulent motions and episodes of intense convection – that can generate modifications within the dynamics of larger scales.

Hence the need for what is known as the 'parameterization of unresolved scales', or the finding of simplified and semi-empirical descriptions of what happens on scales of space and time that the model is unable to reproduce in an explicit way, because they are smaller than the distance between two nodes of the grid used to discretize the equations. This need, which has been well documented for years by those trying to numerically simulate turbulent flows, becomes crucial in climate modelling. In short, the challenge is to find a statistical representation of what happens on the scales that remain unresolved using the simulated behaviour over larger scales, integrating basic physical and chemical principles with empirical expressions derived from observations and experiments. And in parallel to this, it is also necessary to develop implicit descriptions of how the unresolved scales can influence the dynamics of the larger scales. We cannot simulate the orientation of every tree's leaves within a forest, but we can attempt an average description based on large-scale windspeed and attempt to parameterize the mean effect of the ensemble of leaves on the wind.

It is a fascinating and extremely difficult problem that touches the very heart of our ability to represent the world with a mathematical model. If, in order to describe the global climate, we really had to describe in detail each individual leaf or the lightest puff of wind or the details of the waves that break on a beach, then there would be no hope of success. The dynamics of the clouds, for example, covers scales that go from the dimensions of the tiny water droplets to the continental scales, and the distribution of the water droplets can influence the albedo and the greenhouse effect on a planetary scale. Luckily, however, it is not the behaviour of the single drop that counts, but the effect of the entirety of cloud droplet distribution, as discussed in detail in a 2017 article by US climatologist Tapio Schneider and his co-authors. The point is to find a satisfying description of the cumulative effect of the individual components, be it the drops in the clouds or the leaves in a forest or the waves on the sea.

The results of the simulations demonstrate that, when we manage to find proper approximate descriptions of the overall effect of the unresolved scales, then we are able to say something that makes sense. Mostly because, luckily, in many cases information flows from large to small scales, at least in a statistical sense. But this does not always happen. Understanding when this approach is failing and what to do in these cases in order to obtain more correct descriptions is one of the great challenges facing climate modelling and the study of complex systems in general.

Parameterizing complex phenomena on scales smaller than the distance between two nodes on a numerical grid is not, however, the only problem. For some crucial components of the climate system, such as the biosphere, we have seen that the equations of the mathematical models that describe them are substantially empirical and mix basic principles with information that comes from the statistical analysis of data. Once again, we have the issue of parameterizing dynamics that are not described exactly, not only because they occur on too small a scale but also because, unlike what happens with fluid motions, the equations that govern them are unknown. In the study of gases, the dynamics of single molecules is lost in the collective behaviour, described by statistical mechanics and summed up by the laws of thermodynamics. This paradigm also inspires us in the parameterization of the unresolved or unknown dynamics of the Earth System, but we must verify whether or not this is an effectively viable approach case by case.

A Regional Zoom

The resolution of global climate models is often insufficient when it comes to describing what happens in mountain regions, or simulating the dynamics of aquifers, rivers or ecosystems. In all of these cases, we need a finer spatial resolution.

For this reason, in parallel with the construction of global models, over the last 30 years regional climate models have been developed. These are substantially similar in their complexity of description to global models but they cover only a portion of the Earth's surface. There are models for Europe, Africa, Asia, Australia, North and South America and for smaller regions. These models are generally 'nested' within the global ones. Or, rather, the global simulations provide the values on the edge of the region of interest, but, within it, a more detailed description with much higher spatial (and temporal) resolution is activated. Such models dedicated to limited regions can today easily reach resolutions of 10 kilometres, and some push towards 1–2 kilometres, managing to represent much better the effects of orography, the different characteristics of the Earth's surface and the variability of the water cycle.

One of the pioneers of the development and use of regional models is climatologist Filippo Giorgi, whom we have already mentioned. In 2019, Giorgi published an article that takes stock of the situation with regional climate models and discusses possible future developments. Regional models have a research programme dedicated to comparing the results and collecting available simulations. This is CORDEX (Coordinated

Regional Climate Downscaling Experiment), an international initiative for coordinating efforts to refine climate information through dynamic downscaling, which is essentially the nesting of high-resolution regional numerical models within global climate simulations.

Imperfect Models

As we have already said, no model is perfect and no model is an exact reproduction of reality (an old aphorism, attributed to statistician George Box, is that 'all models are wrong, some are useful'). This brings us to an issue that is crucial yet hard to accept, and that is somewhat disliked by the users of science: the uncertainty of results.

By its very nature, science is imperfect. Research provides us with an image of the world and its interconnections that best represents what we have managed to understand, but it is not an absolute truth. It is always only ever a representation – useful, logical, coherent (albeit only partially, as Gödel's theorem tells us), but destined sooner or later to be abandoned in favour of a more articulated version better equipped to explain the new observations that slowly broaden our horizons. Galilean relativity is not 'wrong', but it has limits to its application which were overcome by Einstein's relativity, which expands on them and includes them as a particular case. Classical physics works well for falling apples and planets, but must be expanded to quantum mechanics if we want to describe phenomena on an atomic scale (and maybe even life processes, as developments in quantum biology seem to suggest).

The situation is the same when it comes to the climate. We will never be able to describe 'everything' and must therefore choose a priori the level of detail we want to reach. Furthermore, we are dealing with chaotic, turbulent systems that can easily behave in irregular and unpredictable ways. All of this means that a climate simulation, even when obtained using a deterministic model that resolves well-known and defined equations, cannot be taken as the absolute truth. Rather, it is a possible behaviour of a virtual system that resembles as far as possible the Earth's climate. But the temperature simulated in a specific grid point will never be the 'real' temperature that we can measure at that point and in that specific instant. To all intents and purposes, it is exclusively a statistical indication.

233

This happens because the equations are approximate (equations are not reality), and because our numerical algorithms do not exactly resolve the equations we have chosen. And because we do not know in detail the external forces that act on the system, now or in the future. But also because, even if we had a 'perfect model', small variations in the state of the climate at the beginning of the simulation generally lead to unpredictable changes in the sequence of values that the climate variables will assume. The small book *Chaos: A Very Short Introduction* by mathematician Lenny Smith provides a starting point from which to understand chaotic phenomena, and the book *Reductionism, Emergence and Levels of Reality* by Italian mathematicians and physicists Sergio Chibbaro, Lamberto Rondoni and Angelo Vulpiani (the latter we have spoken about already when discussing stochastic resonance) explores many conceptual issues with the deterministic approach to the description of reality.

Living with Uncertainty

A climate model, though imperfect, is not useless. Because, while it is true that it will not allow us to simulate in detail whether or not it rained in Paris on 19 October 1687, or whether it will be sunny in New York on 24 November 2058, it nevertheless provides reliable information (if we have not made any mistakes when writing and solving the equations) on the statistics of the quantity we are interested in, and on the probability that it could take on certain values. Probability. This is the dragon that anyone taking decisions must face. Scientists are often asked 'What will happen?' by politicians, managers or decision-makers hoping for an unequivocal response: 'This is what will happen.' The scientist's word. And yet, no. In the majority of cases, a serious researcher can only respond that there is a certain probability that something will happen, and a certain probability that something else will happen instead. Obviously, the challenge is to make these probabilities increasingly accurate and, as far as is possible, identify with ever lower uncertainty the situation that is most likely to happen. But it will never be a certainty.

In order to practically determine these probabilities in the case of climate simulations, the most common method is to carry out many simulations, both with slightly varying models (for example, based on different numerical algorithms or modified parameterizations), and using the same model but starting with slightly different initial conditions that generate climate stories with dissimilar details but similar statistics. Bringing together all of these climate stories generates an ensemble of simulations that informs us about the values expected for the climate variables, and on the uncertainty characterizing the estimation of these values. An ensemble can be formed by many simulations obtained using the same model (a multi-member ensemble) or by bringing together the results of many different models (multi-model ensemble).

Figure 18.2, reworked from NASA climatologist Gavin Schmidt's blog, Realclimate.org, takes as an example the estimate of global temperature from 1980, comparing the observational data available with an ensemble of simulations from climate models taken from the CMIP5 archive. The

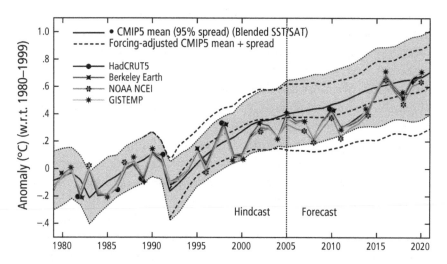

Figure 18.2 Global temperature obtained from four archives of observational data interpolated on a regular grid (lines with symbols) and from an ensemble of climate simulations from the CMIP5 archive (grey band and black curved line). The dashed lines represent the same modelling results using different kind of forcings, as discussed by Gavin Schmidt et al. (2014). For each year, the width of the band includes 95% of the values obtained through simulations. The temperatures are represented as anomalies (differences) with respect to the average temperature in the period 1980–99. Reworked from Schmidt's site, Realclimate. org, originally based on data from the work of Kevin Cowtan et al. (2015).

grey area represents the band of values obtained by bringing together all the model simulations calibrated in the period up to 2005, providing an estimate of uncertainty. The values from simulations beyond 2005 represent a climate projection, an extrapolation of the model beyond the period in which the parameterizations have been calibrated (we will return to climate projections in the next chapter). As we can see from the illustration, the measured data is substantially included in the uncertainty bars of the ensemble of simulations, indicating that we have not gone too wrong in our description of the climate. This brings us to the realm of climate model validation.

Validating the Models

The uncertainty intrinsic in climate models (as in those of any complex system) should not be taken as a sign of pure ignorance. Of course, the more we know, the smaller the uncertainty, but in the majority of cases we cannot eliminate it entirely because it is an essential trait of natural systems. Uncertainty is something very different from possible errors in formulation. We must therefore ensure that our simulations are capable of correctly reproducing what is measured, including uncertainty, in the climate or any other field.

Figure 18.2 provides us with an example of how the ensemble of CMIP5 climate models reproduces measured data, at least on the level of average global temperature. And it also shows us the limits, or rather the breadth, of the uncertainty bar. A similar exercise has now also been carried out for CMIP6, with analogous results. Various comparative works between data and models carried out over the last 30 years confirm that we are able to simulate at least the most general traits of the climate over the last few centuries. However, we said earlier that the models are characterized by semi-empirical parameterizations constructed precisely based on the available data. Therefore, the capacity to reproduce that same data used to create the model is a minimum requisite, which tells us that we are capable of building a self-consistent description of the climate system that includes the more relevant control factors.

Over recent years, climate simulations of the last 100 years have been carried out, both including all types of forcings (including the anthropic ones, linked to the rise in atmospheric CO_2 concentration), and excluding these and using only natural forcings such as solar variability and volcanoes. The results show that, if we do not include anthropic drivers, there is no way of reproducing the rise in temperatures observed since the 1950s. This is yet more evidence of the human role in global warming, as underlined by the work of climatologist Gerald Meehl and his co-authors in 2004, and later refined by many other researchers and included in the IPCC reports.

An even more stringent request is that the model be capable of simulating the climate beyond the period used to calibrate its parameterizations. This verification has been carried out many times, starting with the first climate simulations. Once again, figure 18.2 shows the results obtained for the years following 2005, which represent climate projections beyond the period used to calibrate the model. In this case also, the ensemble captures the observational data, perhaps with a few more uncertainties, but substantially in a similar way to what happens in the calibration period. The 2019 work of US climatologist Zeke Hausfather and his colleagues compares the projections provided by various generations of climate models, verifying their capacity to reproduce observations also beyond the period on which they were calibrated. The models therefore manage to provide important and substantially reliable information on the climate of our planet.

But the Road Is Not Yet Finished

As always, not everything is resolved and we cannot just make do with what we have obtained thus far. For example, when we go to evaluate the most 'difficult' and least homogeneous variables, such as precipitation, soil moisture or evapotranspiration, we have greater discrepancies between the results of the simulations and the observational data. Similarly, if we consider increasingly small scales, particularly in mountain or coastal areas, not even regional climate models are capable of providing entirely satisfying results. The greater the detail we need and the more complex the climate parameters in which we are interested, the greater the uncertainty and, at times, the clear discrepancy between data and models.

A similar problem emerges with the reproduction of the climate from the geological past for periods with climate conditions that were very different from those we have now. For example, to date we have not managed to simulate completely the very hot climate of the Eocene without utilizing values of greenhouse gas concentration that are much greater than those presumably present at that time. It is possible that some parameterizations of the models, necessarily calibrated on the present or the near past, are not suited to describing climates so very different from the one we have today. It is one thing to reproduce a temperature rise of 1–2 °C, but it's another to represent a world that is 10 °C hotter or colder. This aspect sets off alarm bells with regard to the capacity of climate models to represent sudden climate transitions, such as one caused by a widespread release of methane through the clathrate or permafrost destabilization. This is one of the hot topics in climate research, and a great deal of energy is now being spent on increasing the models' capacity to appropriately simulate sudden climate changes.

Another important point is that we have almost always talked about anomalies, the differences between the value of the climate variable at a given moment and its average over a reference period. But the difference between two variables tells us nothing about their absolute value: 102 less 100 has the same difference as 152 less 150. This approach does not allow us to verify whether the model provides absolute values – for example, in degrees Kelvin for temperature – that are comparable with those measured. And it is not actually rare that, in the averages, there are pronounced differences between the different models and between models and observation. The models often generate worlds that are a little hotter or cooler than our own. This means that the difference between the average temperature in 2050 and that in 2000 is a reliable estimate, but the absolute value of the average temperature in 2000 can be out by as much as 2 °C. If it counts only the difference in temperature between the future moment that interests us and the reference period, then there is no problem. But if the effective value of temperature matters (for example, in order to determine the absolute altitude of the zero degree isotherm or the quantity of water vapour that can be contained in the atmosphere), then these discrepancies can become relevant.

Uncertainty cannot be eliminated but it must be reduced as much as possible. Ironically, it would be easy to say that the ensemble of simulations includes the observations if the interval of uncertainty went from absolute zero to 1000 °C! What is needed is to reduce as much as possible the spread of values provided by the ensemble of simulations, continuing nevertheless to reproduce the observations within the uncertainty bars. This is no mean feat, but it is a task to which many research groups and centres the world over have devoted themselves. Climate models are undoubtedly imperfect tools but they are the only ones we currently have to represent the climate of our planet.

The Virtual Earth

In recent years, the concept of climate modelling has expanded towards a new frontier: to build the digital twin of the Earth System. This idea is supported, for example, by the ESA with the Digital Twin Earth project, and by the European Community with the Destination Earth (DestinE) programme. A 2021 article written by European climatologists Peter Bauer, Bjorn Stevens and Wilco Hazeleger, all experts in climate modelling, tackles this revolution in our approach to the study of the climate in an intriguing and articulated way, linking it to the 'ecological transition' that we are trying to undertake on a global scale.

Its starting point are the digital twins of the great industrial plants, made up of a collection of mechanical, chemical and electronic components. They are complicated systems and if we want to know what will happen if a valve breaks or an electrical circuit burns, we cannot simply find it, hit it with a hammer and see what happens. We must have a twin system on which to intervene and evaluate the response. In NASA's Apollo missions, a physical twin was built of the lunar excursion module (LEM), which was identical to the space twin but kept on the base, and on which it was possible to reproduce malfunctions and breakages that could happen to the one on the Moon. An analogue twin.

With the progress of numerical simulations, calculation capacity and Artificial Intelligence techniques, analogue twins were replaced with digital ones, which we can use to predict and cure the possible malfunctions in the 'real' ones. Given, of course, that we have correctly reproduced the way in which the real twin works.

In reality, the Earth's climate is much more complex than any industrial plant. There are enormous quantities of interactions, connections, feedbacks, cycles that take place on temporal scales ranging from fractions of seconds to millions of years, and on spatial scales from a few microns to the entire globe. This is a visionary and epochal challenge that will lead to the development of new generations of models, based on deterministic algorithms and empirical parameterizations; on the continual flow and assimilation of data measured by the sensor networks throughout the land, sea, atmosphere and space; on the real-time management of enormous quantities of data; and on the use of the most advanced Artificial Intelligence techniques. One step at a time, digital twins of specific sub-systems are being developed for a particular ecosystem, a certain area of the sea or strip of soil, with the hope of one day arriving at a digital description of the entire planet. Who knows whether we will actually achieve this, but it is, in any case, a powerful stimulus for further developing our imagination and capacity for understanding the world.

19

Knowing in Order to Anticipate, Anticipating in Order to Act

Everything we have learned about the Earth's climate and climate simulations now accompanies us towards the conclusion of our efforts: estimating what could happen in the future and organizing the measures that will reduce the risks of climate change. Knowing to predict, and then to act, to build on the vision of Auguste Comte.

Meteorology and Climate

Firstly, we must clarify the distinction between meteorological forecasts and climate projections. US mathematician and meteorologist Edward Lorenz, whom we have already referenced and who in 1963 began the numerical study of systems with chaotic behaviour, distinguished between prediction of the first kind, where we want to know the 'exact' state of the system starting from its current conditions (as in weather predictions), and of the second kind, where we want to know the statistics of the system given certain external driving forces (as in climate projections).

In the case of meteorology, with the starting point of a detailed knowledge of the current conditions of the atmosphere, oceans and Earth's surface, our aim is to predict the state of the atmosphere in a few days' time with the greatest possible accuracy. We want to know if and how much it will rain in a certain valley or particular city at a precise moment over the next few days. These forecasts improve as our knowledge of today's meteorological state increases, as the grid of the model we are using to represent the atmosphere grows more refined and our formulations regarding the description of atmospheric motions become more exact, including the parameterizations we discussed in the previous chapter.

The first attempts to obtain quantitative meteorological forecasts were described in 1922 in a pioneering book by British mathematician and meteorologist Lewis Fry Richardson. Here we find the idea of discretization, the grid on which to calculate the equations, and the need to resolve the entirety of the necessary mathematical relations one piece at a time in order to carry out these forecasts. It was still too early, however, and these first approaches were unsuccessful because the equations chosen were too complicated to resolve using the technological means of the era. Some 30 years later, Jule Charney, whom we have already met while talking about deserts, derived a simplification of the equations used by Richardson, managing to resolve them with the computers available at the time at the Los Alamos National Laboratory in the United States. With this combination of mathematical analysis and computing power, the age of quantitative meteorological prediction had begun.

Meteorological forecasts are substantially what is known as an initial value problem, meaning we want to determine the future state using our knowledge of the present one. In the words of Edward Lorenz, a prediction of the first kind. However, it is precisely Lorenz's work that shows it is not possible to predict exactly the atmospheric conditions in any future moment. Indeed, even if we were able to solve the exact equations, the chaotic and turbulent nature of atmospheric motions would amplify any tiny uncertainty in the knowledge of the initial state. On the whole, the limit to meteorological predictability depends on the detail we want to achieve, but today it is estimated that this does not go beyond a week or 10 days for the largest scales, and is much less as a much greater level of spatial detail is required. The improvement of the observational network obviously helps, but the limit is particularly rigorous and it will not be simple – or even possible – to overcome it.

It is natural, then, that we might ask ourselves how we can 'predict' the climate in 50 years? The fact is that, when it comes to the climate, we are never looking for a detailed prediction, but an estimate of the most probable conditions (and their uncertainty) for a particular configuration of the forcing factors. Or a 'projection', described by Lorenz as a prediction of the second kind. The question is what the climate statistics will be in a particular region and in a certain period of time, once the main drivers have been established: solar luminosity, greenhouse gas concentration, surface characteristics and so on. But always only ever in a statistical sense. With these types of probabilistic projections, we can overcome the limit of predictability for detailed meteorological forecasts, and we are able to explore how the climate will respond to natural changes in the drivers or those caused by human activity.

Climate Scenarios

Over the last century, the concentration of atmospheric carbon dioxide has grown due to emissions of anthropic origin. Therefore, in order to evaluate the extent of the change, we must first estimate how much CO_2 will be emitted by human activities in the near future. This gives rise to the concept of 'scenario' because no one can know exactly how society will develop in the coming decades or whether – and when – a transition towards lower greenhouse gas emissions will be put in place. We can, however, identify certain main categories: a continuation of emissions with no effort made to moderate them (business as usual or BAU, also known as 'worst-case scenario'), a slight reduction, a drastic reduction or total cessation. Similarly, we can hypothesize an intense deforestation of the Amazon, as is sadly often threatened, a reforestation of northern mid-latitudes or some other modification in land use. What will happen depends on geopolitical conditions, on development strategies, on what rapidly industrializing countries and those that are fully industrialized will do. What researchers can do is identify a collection of possible scenarios for socio-economic evolution, associating them with different configurations of climate forcing factors.

In recent years, these scenarios have been expressed in terms of the 'radiative forcing' equivalent to a certain concentration of greenhouse gas. This refers to how many additional watts per square metre will be generated by all of the anthropic forcing in 2100, as discussed in the IPCC 2007 Expert Report, edited by climatologist Richard Moss and his colleagues, and later extended to include changes in socio-economic development and land use, as described in 2017 by the expert in energy and climate Keywan Riahi, based at the International Institute for Applied Systems Analysis (IIASA) in Austria, and his co-authors. The possible scenarios are referred to as Shared Socio-economic Pathways (SSP) – that is, possible paths that represent changes in greenhouse gas emissions and socio-economic conditions. Each scenario is characterized by a prime number from 1 to 5, indicating the typology of development, and a second number that indicates the equivalent radiative forcing in

2100. In this way, we find the worst-case scenario, SSP5-8.5, in which emissions continue without any decrease, leading to an increase of 8.5 W/m² by the end of the century; or an intermediary scenario, SSP2-4.5, with modest reduction in emissions and an additional 4.5 W/m² expected by 2100; or a virtuous scenario, SSP1-1.9, in which emissions are cut drastically in the immediate future.

Each of these scenarios for emissions and socio-economic development is used to force an ensemble of global climate models, obtaining an estimate of how the climate system will respond over the coming decades. As always, different models provide slightly different responses, and at the end we will have a collection of possibilities characterized by an average response accompanied by a specific range of uncertainty. Figure 19.1 shows the projections for global temperature and decrease in Arctic marine ice in the different scenarios obtained from all the models used in the IPCC's sixth report. As we can see, while a modest rise in temperature is expected in scenario SSP1-1.9, even followed by a slight decrease before 2100, in the worst-case scenario, SSP5-8.5, there could be an average increase of almost 5 °C compared to the period 1850–1900 (taken as reference). The Arctic will continue to warm up more than the rest of the world, and, in the worst scenarios, a total disappearance of late-summer marine ice is expected. In these same scenarios, a notable increase in temperature is also expected in the Antarctic region.

The projections for precipitation are particularly important, even if they are still affected by notable uncertainty. In the scenario with the greatest emissions (SSP5-8.5), a significant increase in precipitation is expected in both equatorial and polar regions, where the rise in temperatures and the disappearance of marine ice could intensify the water cycle. In all of these scenarios, however, a decrease in precipitation is expected in southern Europe and in the south-west of the North American continent, amplifying the tendency towards drought that we have already witnessed in recent decades.

Unexpected Instabilities

In the majority of cases, climate projections indicate a continuation of the behaviour we can already measure today, more or less accelerated depending on the pace of greenhouse gas emissions. However, the

a) Global surface temperature change relative to 1850-1900

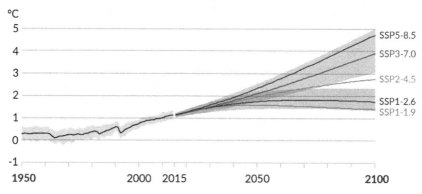

b) September Arctic sea ice area

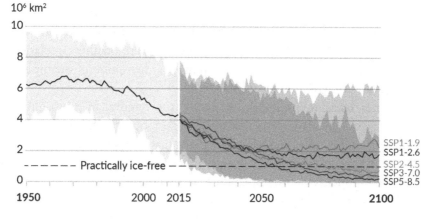

Figure 19.1 Expected growth in average global temperature (top) and decrease in Arctic marine ice cover at the end of summer (bottom) for the various SSP climate scenarios. The temperature values shown in the top panel are differences from the reference period 1850–1900. The top and bottom curves represent the two scenarios SSP5-8.5 (the 'worst case') and SSP1-1.9 (the 'virtuous case'). The other scenarios are intermediate. The shaded areas represent the uncertainty estimated from the ensemble of climate simulations. Taken from the Summary for Policy Makers in the sixth IPCC report (AR6), www.ipcc.ch/assessment-report/ar6.

climate does not always behave in a predictable way due to the strongly non-linear processes that characterize it. Some mechanisms in particular can act as amplifiers of global changes.

For example, if there were a rapid release of methane produced by the extensive decomposition of organic matter contained in the permafrost, or by the instability of clathrates, there could be a sudden increase in the greenhouse effect, with a rise in temperature well above those we have currently predicted. This is a bit like what we believe happened between the Paleocene and the Eocene during the thermal maximum some 55 million years ago.

In the Arctic seas, the disappearance of marine ice strengthens the heat exchange between ocean waters and the atmosphere, leading to modifications in the hydrological cycle and an increase in cloud cover, with effects that could proliferate throughout the entire climate system. Furthermore, the melting of the ice sheets releases fresh water into the sea, with a possible subsequent slowing of the thermohaline circulation leading to its collapse, or at least to a powerful disruption of the bipolar oscillation with global consequences.

Another relevant mechanism concerns the removal of atmospheric carbon dioxide by plankton. Put simply, while zooplankton feed on other organisms and extract energy from organic material, phytoplankton use photosynthesis to absorb CO_2 and directly produce organic matter. Some components of phytoplankton (coccolithophores, for example) or zooplankton (like some foraminifera) build shells from calcium carbonate. After the death of the organism that built them, many of these shells sink and are deposited in the sediment. The carbon that was originally absorbed through photosynthesis is then stored at the bottom of the sea, helping to keep the rise in CO_2 concentration under control.

However, the rise in atmospheric carbon dioxide also causes a rise in the CO_2 dissolved in the ocean waters. When dissolved CO_2 increases, the water becomes more acidic (its pH decreases), a fact that can generate problems for many marine organisms. In particular, more acidic waters make it more difficult for plankton to form solid shells, potentially causing a decrease in carbon deposits on the sea bed. If the acidity of the oceans were to increase too much, therefore, the entire removal process could slow, leading to a rise in the concentration of atmospheric carbon dioxide and amplifying global warming. In reality, this entire process is extremely complex and it is not yet clear what the marine ecosystem's global response will be to the combined rise in acidity and temperature, as discussed by oceanographer Jan Taucher and colleagues in 2021. As ever, the ecosystems have extremely complex responses to environmental changes, responses which are not always simple to estimate a priori and which are sometimes counter-intuitive.

The conditions of potential instability are called tipping points, climate situations in which there is a sudden change from one state (for example, the current thermohaline circulation) to a very different one (such as this circulation collapsing). They are not necessarily transitions that are irreversible in the long term, but they can have significant consequences for hundreds or thousands of years. We do not yet know enough about these sudden transitions, nor are we able to recreate them with sufficient precision using models. There is still a lot to do, but the issue of tipping points is extremely relevant from both a scientific and a practical point of view. In 2019, UK climatologist Tim Lenton and a sizeable number of co-authors published an important discussion with a very evocative title, in which they examined the risks linked to the triggering of these potential instabilities – 'Climate tipping points: too risky to bet against'.

Estimating the Impacts of Climate Change

Knowledge of climatic variables such as temperature, precipitation, wind or marine currents is an indispensable starting point, but it is not enough. Often, we are more interested in knowing how the variation in climate will influence the recharge of aquifers, or the probable extent of the area burned by forest fires, or agricultural productivity. Or, again, how biodiversity or the way in which ecosystems function might be modified by different climatic conditions, or how the glaciers in a certain mountain region will respond. Or we are interested in the impacts of climate change on the land and the environment. In all of these cases, climate science must dialogue with geomorphology, with ecology, with hydrology. Similarly, we might be interested in how climate change will influence society, the economy, health, interacting for example with the effects of atmospheric pollution or facilitating the diffusion of insects carrying pathologies linked to tropical fevers.

As an example of a study on climate impact, figure 19.2 shows the projection of the average retreat of the front of fourteen glaciers in the Italian north-western Alps for the so-called RCP (Representative Concentration Pathway) 8.5 scenario, similar to SSP5-8.5. The analysis of observational data collected by the Italian Glaciological Committee allowed for the construction of an empirical model for the average response of glaciers to changes in winter precipitation and summer temperatures. The empirical model for glaciers, once validated, was forced with the results of an ensemble of simulations carried out using the European global climate model EC-Earth. As we can see, the expected glacier retreat in this scenario becomes increasingly rapid over time, leading to the potential disappearance of various Alpine glaciers within a few decades. Similar projections have been carried out for individual glaciers or groups of glaciers in different parts of the world.

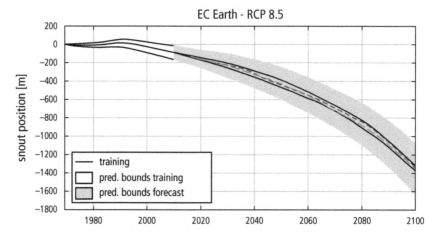

Figure 19.2 Projection of the average retreat of the fronts (snouts) of fourteen glaciers in the north-western Italian Alps. The frontal position is measured by the difference in comparison to that in 1968. The observational data is in the white part and is used to build a model of the glacier response. The curves after 2010 represent the average from the projections obtained using various simulations of the global climate model EC-Earth for the RCP 8.5 scenario. The shaded area measures the range of uncertainty in the results, obtained by combining the uncertainties from the various global simulations and from several realizations of the empirical model for glacial response (taken from Bonanno et al., 2013).

The IPCC reports are overflowing with examples of how various environmental systems can respond to the global warming predicted by various emission scenarios. In all of these cases, the output of climate simulations, be they global or regional, are used as weather and climate forcing for models of specific environmental systems. For example, we will use a mathematical description of an ecosystem in a very precise area of study that will then be submitted to future conditions simulated by the climate model. This means we will build a 'digital twin' of the ecosystem under examination and explore what happens to it if the forcing to which it is subjected changes. The results of the simulations will indicate how a particular environmental system will be able to react to variations in the climate, but also how it will respond to possible management actions to reduce the damage and to environmental restoration interventions. Particular attention will be given to the possible loss of 'ecosystem services' following climate change, meaning all of those natural benefits that functional ecosystems provide to humanity, such as water and clean air, the stability of mountainsides, control of the water cycle and many, many more.

Levels of Risk

In practice, we do not know whether the countries of the world will manage to come together for an effective reduction in emissions or whether everything will continue as it has until now. In this sense, the instabilities and sudden changes in global politics, for better or worse, must be added to climate factors. Furthermore, it is not only greenhouse gases that count, but the release of atmospheric particulate, changes in land use and possible actions to recover and store excess CO_2. This collection of factors, which is uncertain because it is entirely dependent on geopolitical dynamics, suggests that we should not only focus on the response in a specific scenario but, rather, try to understand how the environment will react to a given rise in global temperature.

In this approach, we first define the level of the rise in temperature that we intend to explore, moving then to using all the available simulation results that correspond to the pre-chosen level of temperature increase, independently of when they might happen or in which specific scenario. For example, if we are interested in understanding what might happen to a particular ecosystem with a rise in temperature of 3 °C, we will use both the results of scenario SSP5-8.5 predicted in the next few decades, and those for scenario SSP3-4.5 predicted for the end of the century, bringing together all the possible climate conditions associated with such an increase in global temperature in order to see the effect they produce.

As a specific example of this approach, we will consider the problem of forest fires in Mediterranean Europe. Figure 19.3 shows the risk of an increase in the burned area for various levels of warming, as obtained by climatologist Marco Turco and his colleagues using an empirical fire model forced by global climate simulations. The more the temperature rises, the more the combination of scarce precipitation and greater evapotranspiration will favour the development of more extended wildfires. The current measures for prevention and control, which have allowed for a decrease in the burned area over the last 30 years, would not be capable of effectively controlling the fire risk and would have to be significantly reinforced. Similar analyses can (or should) be carried out for all regions of the world at risk of forest fires.

Similar approaches can be used to estimate the impacts of climate change on every other kind of environmental system. With this in mind, in 2018 the IPCC published an extremely important special report titled *Global Warming of 1.5 °C*, which further reduces the maximum increase of the global average temperature at which it is less difficult to contain the damage. Figure 19.4, taken from the special IPCC report cited above, shows a series of impacts on specific systems expected at different levels of warming. The darker the column, the greater the risk of damage.

As we can see, not all systems respond in the same way. Tropical coral is already now showing serious damage, just as the risk of coastal flooding is already very high. The small fisheries at low latitudes and the Arctic regions are in danger with every further rise in temperature. The mangroves and tourism, however, do not seem to be particularly affected by global warming. It was this work that showed how the previously identified limit of a temperature rise of 2 °C when compared to the pre-industrial period is overly lax. In order to avoid negative effects that are difficult to contain, it will be necessary to stay below a 1.5 °C rise in temperature compared to the end of the nineteenth century – so, less than 0.5 °C compared with today.

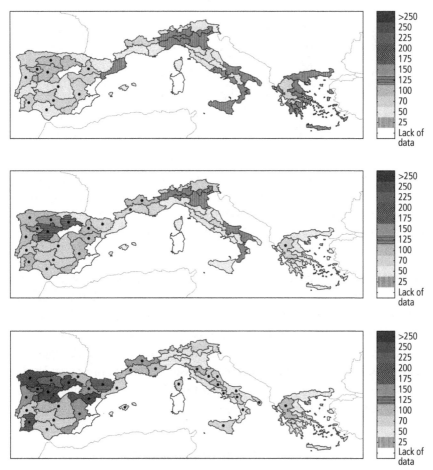

Figure 19.3 Map of the percentage rise in the area burned by summer fires compared to average values from recent decades, for global temperature rises of 1.5 °C (at the top), 2 °C (in the middle) and 3 °C (at the bottom), as compared to pre-industrial conditions, for Mediterranean Europe. The horizontal axes represent longitude and the vertical ones latitude. Taken from Marco Turco et al. (2018).

If we surpass these limits, the world will not end. It is simply that, beyond these values, the costs of 'repairing' – or, in any case, dealing with – the damage caused by climate change become greater than the costs of enacting an effective reduction in greenhouse gases. These limits tell us that, with higher temperatures, some ecosystems could collapse and the environment could be subjected to notable stress. With higher

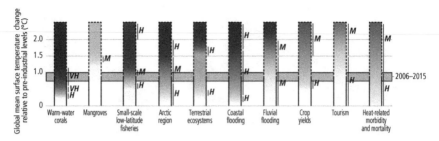

Figure 19.4 Expected impacts of a particular rise in global temperature (indicated on the left). The grey horizontal band indicates the current conditions. The darker the columns, the more serious (qualitatively speaking) the risk. The letters next to the columns indicate the interval of reliability in the results: M: medium, H: high, VH: very high. Extracted from the IPCC Special Report 'Global warming of 1.5 °C' (2018).

temperatures, the world to which we are adapted, in which we have built complex infrastructure, often on the coast, where the human population has become increasingly abundant but more than half of the human population still live in conditions of poverty or insecurity, would become progressively perilous for our societies. Combatting climate changes and respecting the speed limits (of the increase in greenhouse gases in the atmosphere) is insurance for our own future. It is an ambitious challenge that requires an immediate response.

Operational Climatology

In order to be of practical use, the results of climate research must leave the purely academic realm and become a source of concrete information for those tasked with managing development strategies, economic decisions, agricultural and industrial production. For this reason, in recent years 'climate services' have been developed, following the example of meteorological services that inform people on what the weather will be like over the next few days.

In the field of climate services, or in what is known more generally as operational climatology, the aim is to provide relevant information for a set of particularly important time frames. The first is the seasonal scale: knowing whether the next summer season will be cool and rainy or whether there will be drought, whether the next winter will be freezing or mild, whether there will be plenty of snow or very little, allows us to provide strategies for the water and energy resources, to plan agricultural and tourist activity, and to prepare for the dangers of fire or flooding.

Clearly, for reasons discussed at the beginning of the chapter, seasonal forecasts are always only ever probabilistic, but they should become a conclusive guide on strategies for adapting to changes in the climate. Seasonal forecasts are difficult, because they exist on the boundary between detailed meteorological forecasts (of the first kind, to use Lorenz's language) and statistical climate projections (of the second kind). They depend on external forcing, but also on the current state of the climate and the interactions between the various components of the Earth System. At the moment, seasonal forecasts are still in their infancy, but given their practical importance, a great deal of work is being done to improve them and make them more reliable.

The longer timeframes are also important, because planning and creating infrastructure takes years and must be functional for a long time. Operational climatology and climate projections on a scale of a few decades are necessary for planning interventions in agriculture, for designing strategies of sustainable development and for using natural resources in an intelligent way so that we might make a real ecological transition guided by knowledge and innovation. Knowing what precipitation will look like over the next 20 years will help us decide where and how to build dams, aqueducts, wind farms – in short, to identify the most opportune transformations of land use.

There will be difficult decisions to make. If we are facing a lack of water in the summer, perhaps due to a reduction in glaciers and snow cover, can we intervene by implementing a system of small artificial mountain reservoirs that release water in the most critical periods, and do we want to? Where do we intend to build large wind farms? How much do we want to intervene in the environment to mitigate the risks associated with a rise in temperature? These are crucial discussions for the coming years. Climate study has now left research centres and entered the construction of our very future. And it is important that this is the case, that scientific knowledge gives rise to forecasting and, from this, conclusive action that aims to marry the well-being of our species with the environmental sustainability of our actions. But this is not always easy.

Reducing the Concentration of Greenhouse Gases

If we want to reduce the rise in temperatures, we must reduce the concentration of greenhouse gases in the atmosphere. Or at least not increase them any further. The challenge, therefore, is to get halfway through this century with a drastic reduction in emissions of carbon dioxide and other greenhouse gases, otherwise the temperature will rise even faster, bringing with it the damage we have discussed. This could be the beginning of a complicated period, in which it may be increasingly difficult to reach an acceptable level of social and geopolitical justice.

Naturally, this is much easier said than done. The conversion of energy production from fossil fuels to sources with low carbon emissions can be costly and complex. But in all countries the world over, willing or not, this substitution is at the centre of people's attention, as well as that of the politicians and the industrialists. Renewable sources (solar, wind, hydroelectric), geothermal, nuclear energy (from fission or fusion, when it happens) are all possibilities that must be discussed and analysed. There is research into artificial photosynthesis and new kinds of batteries capable of accumulating and maintaining energy and feeding it back into the network when necessary. Some paths can be taken, others not yet, others are not appropriate, bearing in mind that a means of energy production (particularly one that provides the enormous quantities of energy we increasingly need) that is entirely harmless to the environment does not exist. We need to evaluate risks and benefits, costs (also in terms of the environment and public health) and advantages – in a secular way and without prejudice, basing our opinions on the facts and results of research. It is no coincidence that the majority of oil companies are diversifying their own interests, moving towards renewable energy and new technologies with low greenhouse gas emissions.

Individual choices are also important, as is awareness and making an effective contribution to a reduction in emissions. Using the car only when it is strictly necessary, choosing public transport where possible; using a bicycle; not heating or cooling houses to the point we have to wear a T-shirt in the winter and a jumper in the summer; eating less meat hailing from intensive farms; informing ourselves on food and the ecological footprint of its production, packaging and distribution; being careful not to waste energy or water. These are all things that help us live better and help us save money, but they do not need to become a kind of forced penance for our sins. They help, but the solution is social and political and it is even more important to demand that governments define and respect large-scale environmental strategies. Not as a return to some lost era, as we have already said, but as a transition towards an even more advanced age, as a motor for the development of new ideas and new technologies, combining well-being, sustainability and the awareness that we are part of nature and of this planet and that we must respect the rules.

Intervening on the Climate

Alongside the reduction in emissions, current research is also concentrating on possible ways of avoiding carbon dioxide being released into the atmosphere, capturing it at source – for example, when it is emitted from large industrial plants – and storing it in the depths of the Earth's crust or even at the bottom of the sea.

Carbon capture and storage (CCS) is an extremely relevant issue and taken very seriously by research bodies, governments and energy-producing companies. The simplest approach is to store carbon dioxide in porous layers beneath impermeable rocks, which stop it from escaping. However, this purely geological form of storage can be difficult due to possible fractures and faults in the rocks above, which would make them less impermeable and could lead to an escape of the 'mechanically' trapped gas below.

Of greater interest is the mineralogical capture of carbon dioxide, which is the possibility of making CO_2 interact with the right rocks in order for it to be absorbed directly by the crystalline structure of the minerals. In Iceland, the research programme CarbFix, financed by the country's Environment Agency and energy companies working with many research centres throughout the world, has the capacity to mineralize carbon dioxide through chemical reactions with the basalt at depths of more than 400 metres. A similar approach, discussed by CNR geochemist Chiara Boschi and her colleagues, is based on the reaction of CO_2 with the silicates and magnesium oxide contained in rocks known as serpentinites, in order to form stable carbonates (magnesite, calcite, dolomite).

Even if it is already in an advanced state of applicability, the technology behind CCS and mineral sequestration clears a path for other problems that require resolving, ranging from the elevated costs involved to the social acceptability of actions of this kind, but it is at the very least necessary to consider all the possible routes. Because it is useful, but not enough, to travel by bicycle rather than by car. The solution will, inevitably, be provided by a complex and articulated collection of industrial initiatives, new technologies, and innovative and diversified ideas.

One final point involves something sometimes referred to as 'geoengineering', the large-scale modification of the Earth's environment in order to reduce global temperatures. Various proposals featuring varying degrees of peculiarity have been made over the years, but to date none has been truly reasonable. The most well known is the idea of filling the upper atmosphere with reflective particles in order to increase the albedo and decrease the amount of solar energy arriving on the planet's surface. Beyond how achievable this project actually is, various modelling simulations have shown the notable risks associated with such operations. In particular, the hydrological cycle could be seriously disturbed, with entire areas of the planet exposed to a higher risk of drought. In short, I believe it's best to avoid playing at being the sorcerer's apprentice, particularly with a system as complex as the Earth's climate. The main route is to reduce emissions, increase the use of energy sources with low carbon content, make efficient use of renewable energies and eventually work on the capture and storage of CO_2 for the time that is strictly necessary to allow a full energy transition.

Who Pays the Bill?

Energy transition is a great opportunity for development, but its initial costs must be covered. So, who is going to take on these expenses? Energy producers or end users, or (as seems right) a balanced division of costs between producers and consumers? At this point, another question arises: who decides what a 'balanced division' looks like?

Furthermore, in the hypothesis that energy transition is not carried out quickly, who will bear the enormous costs of the damage caused by global warming? Typically, these costs are met by governments, which, by using the money that comes from taxation, inevitably burden their citizens.

With this question we enter a delicate social and economic realm with even vaster implications for the consequences of climate change. It concerns how economic resources are distributed among a population and the different nations. The Gini Index measures the distribution of resources among a population (of individuals or states). This index goes from 0, where the resources are distributed in an equal way among all components of the group, and 100, where all the resources are concentrated in the hands of one single member of the group.

In the passage from a hunter-gatherer society to that of non-migratory farmers, it is estimated that the Gini Index went from around 20 to more than 40 during the Roman Empire. Today, it varies from one nation to the next. For example, estimates made by the World Bank for the years between 2015 and 2019 indicate around 35 for the UK, 41 for the US, 38.5 for China, 37.5 for Russia, around 32 for Germany, but more than 63 for South Africa (from 2014). An even more worrying piece of data is that we can see that there is a tendency almost everywhere towards a rise in resources being concentrated in very few hands, with a small number of individuals increasing their percentual incomes much more than the vast majority of the population. It seems, therefore, that we are witnessing a possible transformation towards a society in which the middle classes, a driver of progress since the Middle Ages, could progressively lose their relevance, crushed between a much reduced group that controls the majority of resources and an almost totality of the population in precarious economic conditions. And this is a society that could quickly become unstable.

This increasingly unequal distribution of economic resources has important consequences, because the availability of resources brings with it the capacity for decision-making. This leads us back to the problem of who chooses which path we will take. If the decisions about our future are taken exclusively by a highly reduced group of people (no matter how capable and honest they might be), we risk a dystopian society that we would prefer to see only in science-fiction films, with a crazed climate and desperate masses who are barely surviving in an authoritarian society controlled by a few individuals separate from the rest of humanity. It is up to us, then, to act so that this unhappy possibility does not become reality.

Conclusion: The Journey Continues

Until the sun becomes a 'red giant', the Earth will presumably continue to host life. *Homo sapiens* will evolve – perhaps they will be suddenly substituted – but almost certainly, in some 10 million years, the world will no longer be as it is now. There will be new organisms; evolution will continue along the paths dictated by chance and necessity. Other beings will appear, perhaps more intelligent than us, perhaps not. According to James Lovelock, as he writes in his last book *Novacene*, we could be substituted by robotic beings capable of reproducing and able to understand the world better than us. Perhaps. Science fiction offers us a wide range of possibilities, both wonderful and terrifying.

In the meantime, however, during our time on this planet, we must do all that is necessary for the well-being of our species and all passengers on Earth and stick to the rules of the game. Respecting the rules is fundamental, because nature is inescapable. If we do not play by its rules, sooner or later we will be brushed aside. But in order to respect the rules, we must know them. We must understand how things work. This book has tried, in its own small way, to provide some useful information with which to gain a better understanding of our planet's climate.

During our exploration, we have gone from the fearful Hadean era, when the planet's surface bubbled with magma, to the global warming of the last century. We have seen the Earth transformed into a snowball and the high temperatures of the Eocene, along with many other important moments in the history of the climate. We have learned that the climate varies over time and space, sometimes significantly. We have attempted to reveal its mechanisms, the complex interactions, the feedbacks, the dynamics of the oceans, winds and clouds. What happens in the soil, in the Earth's interior and in the biosphere. It is a complex system in constant evolution, perennially changing, but also capable of hosting life, which has in turn influenced it, modified it, amplifying or mitigating its fluctuations. Ours is a living planet that should be loved and respected, without us ever forgetting that it could destroy us at any moment.

We have learnt that one of the main mechanisms for controlling the climate has always been the concentration of greenhouse gases in the atmosphere. It is not the only one but it is very important. Over the last century, human activity has joined the natural variability of the climate, leading to the emission of the largest quantities of carbon dioxide that have accumulated in the atmosphere and the oceans in recent times. As a consequence, the temperatures have increased at perhaps the highest rate ever seen. Climate models, which attempt to reproduce how the Earth System functions, show us that the temperature will increase further, and the more CO_2 is emitted, the faster this will happen. Alongside this, we have also destroyed enormous quantities of tropical forest, replacing them with expanses of the same plants, such as oil palms in Borneo. We have caused the desertification of the oceans, almost exterminated the whales, destroyed the coral reefs. We have dramatically reduced the biodiversity in all ecosystems and eliminated entire marine and terrestrial habitats.

Now we need to come back to our senses, reduce emissions drastically, re-naturalize the environment, stop the degradation of the soil and the extermination of and in the oceans. We must develop strategies for adapting – intervene with new ideas, new technologies, new approaches, new lifestyles. We must adopt systems of energy production that are less primitive and use the power of the sun, the wind and the water. We must build a more modern and more sustainable future, working with nature and not against it, as David Attenborough says in his film *A Life on Our Planet* – developing Nature-Based Solutions, learning to use the forces of nature by keeping in harmony with it. The exploration carried out in this book has come to an end, but our real journey – for a knowledge that brings awareness and allows us to make the world a more just place for all – is only just beginning, like that of the now-distant *Voyager* with its golden disc and its message of peace.

Bibliography

1 From the Ocean of Magma to the Great Oxygenation

Canfield, Donald E., 'The early history of atmospheric oxygen: homage to Robert M. Garrels', *Annual Review of Earth and Planetary Sciences*, 33, 2005, pp. 1–36.

Canfield, Donald E., *Oxygen: A Four Billion Year History*, Princeton University Press, 2015.

Egglseder, Mathias S., et al., 'Colloidal origin of microbands in banded iron formations', *Geochemical Perspectives Letters*, 6, 2018, pp. 43–9.

Hazen, Robert M., *The Story of Earth*, New York: Penguin, 2012.

Kasting, James F., and James Catling, 'Evolution of a habitable planet', *Annual Review of Astronomy and Astrophysics*, 41, 2003, pp. 429–63.

Knoll, Andrew, *Life on a Young Planet*, Princeton University Press, 2015.

Miller, Stanley, 'A production of amino acids under possible primitive earth conditions', *Science*, 117, 1953, pp. 528–9.

Sagan, Carl, and George Mullen, 'Earth and Mars: evolution of atmospheres and surface temperatures', *Science*, 177, 1972, pp. 52–6.

Trail, Dustin, Edward B. Watson and Nicholas D. Tailby, 'The oxidation state of Hadean magmas and implications for early Earth's atmosphere', *Nature*, 480, 2011, pp. 79–82.

2 A World of Fire and Ice

Coleman, Arthur P., 'A lower Huronian ice age', *American Journal of Science*, 23, 1907, pp. 187–92.

Harland, Walter B., 'Critical evidence for a great infra-Cambrian glaciation', *International Journal of Earth Sciences*, 54, 1964, pp. 45–61.

Hoffman, Paul F., and Daniel P. Schrag, 'The Snowball Earth hypothesis: testing the limits of global change', *Terra Nova*, 14, 2002, pp. 129–55.

Homann, Martin, et al., 'Microbial life and biogeochemical cycling on land 3,220 million years ago', *Nature Geoscience*, 11, 2018, pp. 665–71.

Hyde, William T., Thomas J. Crowley, Steven K. Baum and William R. Peltier, 'Neoproterozoic "Snowball Earth" simulations with a coupled climate/ice-sheet model, *Nature*, 405, 2000, pp. 425–9.

Kirschvink, Joseph L., 'Late Proterozoic low-latitude global glaciation: the Snowball Earth', in James W. Schopf and Cornelis Klein (eds.), *The Proterozoic Biosphere*, Cambridge University Press, 1992, pp. 51–2.

Knoll, Andrew, *Life on a Young Planet*, Princeton University Press, 2015.

Mawson, Douglas, 'The late Precambrian ice-age and glacial record of the Bibliando Dome', *Journal and Proceedings of the Royal Society of New South Wales*, 82, 1949, pp. 150–74.

Roscoe, S. M., *Huronian Rocks and Uraniferous Conglomerates in the Canadian Shield*, Geological Survey of Canada Paper, 68-40, Ottawa: Geological Survey of Canada, 1968.

Thomson, James, *On the Stratified Rocks of Islay*, Report of the 41st Meeting of the British Association for the Advancement of Science, Edinburgh, London: John Murray, 1871, pp. 110–11.

Young, Grant M., 'Precambrian glacial deposits: their origin, tectonic setting, and key role in Earth evolution', in John Menzies and Jaap J. M. van der Meer (eds.), *Past Glacial Environments*, Amsterdam: Elsevier, 2017.

3 Light Reflected, Light Re-radiated

Arrhenius, Svante, 'On the influence of carbonic acid in the air upon the temperature of the ground', *Philosophical Magazine and Journal of Science*, 41, 1896, pp. 237–76.

Ghil, M., and V. Lucarini, 'The physics of climate variability and climate change', *Reviews of Modern Physics*, 92, 2020, 035002: https://doi.org/10.1103/RevModPhys.92.035002.

Kopp, Gregg, and Judith L. Lean, 'A new, lower value of total solar irradiance: evidence and climate significance, *Geophysical Research Letters*, 38, 2011, L01706.

Newton Foote, Eunice, *Circumstances Affecting the Heat of the Sun's Rays*, report presented to the American Association for the Advancement of Science (AAAS), 1856. See also www.nytimes.com/2020/04/21/obituaries/eunice-foote-overlooked.html.

Solar Radiation and Climate Experiment (SORCE): https://lasp.colorado.edu/home/sorce.

https://earthobservatory.nasa.gov/features/Tyndall.

4 The Explosion that Changed the World

Bobrovskiy, Ilya, et al., 'Ancient steroids establish the Ediacaran fossil Dickinsonia as one of the earliest animals', *Science*, 361, 2018, pp. 1246–9.

Eldredge, Niles and Stephen J. Gould, 'Punctuated equilibria: an alternative to phyletic gradualism', in Thomas J. M. Schopf (ed.), *Models in Paleobiology*, San Francisco: Freeman Cooper, 1972, pp. 82–115.

Falkowski, Paul G., *Life's Engines: How Microbes Made Earth Habitable*, Princeton University Press, 2015.

Glaessner, Martin F., *The Dawn of Animal Life*, Cambridge University Press, 1984.

Gould, Stephen J., *Wonderful Life: The Burgess Shale and the Nature of History*, New York: W. W. Norton & Co., 1989.

Knoll, Andrew, *Life on a Young Planet*, Princeton University Press, 2015.

Kuhn, Thomas, *The Structure of Scientific Revolutions*, University of Chicago Press, 1962.

Lyell, Charles, *Principles of Geology*, London: Murray, 1830–3.

Nursall, J. R., 'Oxygen as a prerequisite to the origin of the Metazoa', *Nature*, 183, 1959, pp. 1170–2.

Seilacher, Adolf, Dmitri Grazhdankin and Anton Legouta, 'Ediacaran biota: the dawn of animal life in the shadow of giant protists', *Paleontological Research*, 7, 2003, pp. 43–54.

www.nature.com/articles/d41586-020-02985-z.

www.scientificamerican.com/article/say-hello-to-dickinsonia-the-animal-kingdoms-newest-and-oldest-member.

5 Between Catastrophes and Opportunities

Alvarez, Luis W., et al., 'Extraterrestrial cause for the Cretaceous–Tertiary Extinction', *Science*, 208, 1980, pp. 1095–1108.

Berger, Wolfgang H., 'Cesare Emiliani (1922–1995), pioneer of ice age studies and oxygen isotope stratigraphy', *Comptes Rendus Palevol*, 6, 2002, pp. 479–87.

Berner, Robert A., 'The rise of plants and their effect on weathering and atmospheric CO_2', *Science*, 276, 1997, pp. 544–6.

Bonneville, Steeve et al., 'Molecular identification of fungi microfossils in a Neoproterozoic shale rock', *Science Advances*, 6, 2020.

Broecker, Wallace S., 'CO_2: Earth's climate driver', *Geochemical Perspectives*, 7, 2018, pp. 117–96.

Chamberlin, Thomas C., 'An attempt to frame a working hypothesis of the cause of glacial periods on an atmospheric basis', *Journal of Geology*, 7, 1899, pp. 545–84.

Faure, Gunter, and Teresa M. Mensing, *Isotopes: Principles and Applications*, Hoboken, NJ: Wiley, 2004.

Hartmann, Jens, 'Plate tectonics, carbon, and climate', *Science*, 364, 2019, pp. 126–7.

Hueber, Francis M., 'Rotted wood–alga–fungus: the history and life of *Prototaxites* Dawson 1859', *Review of Palaeobotany and Palynology*, 116, 2001, pp. 123–58.

Kent, Dennis V., and Giovanni Muttoni, 'Modulation of late Cretaceous and Cenozoic climate by variable drawdown of atmospheric pCO_2 from weathering of basaltic provinces on continents drifting through the Equatorial Humid Belt', *Climate of the Past*, 9, 2013, pp. 525–46.

Lenton, Timothy M. et al., 'First plants cooled the Ordovician', *Nature Geoscience*, 5, 2012, pp. 86–9.

Loron, Corentin L., et al., 'Early fungi from the Proterozoic era in Arctic Canada', *Nature*, 570, 2019, pp. 232–5.

Macdonald, Francis A., et al., 'Arc-continent collisions in the tropics set Earth's climate state', *Science*, 364, 2019, pp. 181–4.

McKenzie, Ryan N., et al., 'Continental arc volcanism as the principal driver of icehouse–greenhouse variability', *Science*, 352, 2016, pp. 444–7.

Royer, Dana L., et al., 'CO_2 as a primary driver of Phanerozoic climate', *GSA Today*, 14, 3, 2004.

Smith, Martin R., 'Cord-forming Palaeozoic fungi in terrestrial assemblages', *Botanical Journal of the Linnean Society*, 180, 2016, pp. 452–60.

Urey, Harold C., 'The thermodynamic properties of isotopic substances', *Journal of the Chemical Society*, 1947, pp. 562–81.

https://microbiologycommunity.nature.com/posts/5320-it-s-a-fungus-oldest-land-fossil-found.
www.nationalgeographic.com/science/prehistoric-world/permian-extinction.
www.nature.com/articles/d41586-019-01629-1.
www.scientificamerican.com/article/first-life-on-land.

6 The Living Planet

Baudena, Mara, Fabio D'Andrea and Antonello Provenzale, 'A model for soil–vegetation–atmosphere interactions in water-limited ecosystems', *Water Resources Research*, 44, W12429, 2008.

Charney, Jule, 'Dynamics of deserts and drought in the Sahel', *Quarterly Journal of the Royal Meteorological Society*, 101, 1975, pp. 193–202.

Crist, Eileen, and Bruce Rinker (eds.), *Gaia in Turmoil*, Cambridge, MA: MIT Press, 2009.

Falkowski, Paul G., *Life's Engines: How Microbes Made Earth Habitable*, Princeton University Press, 2015.

Humboldt, Alexander von, *Kosmos. Entwurf einer physischen Weltbeschreibung*, Stuttgart and Tübingen: Cotta, 1845–62 (English trans.: *Cosmos*, Cambridge University Press, 2010).

Jones, Clive G., J. H. Lawton and Moshe Shachak, 'Organisms as ecosystem engineers', *Oikos*, 69, 1994, pp. 373–86.

Latour, Bruno, and Peter Weibel, *Critical Zones: The Science and Politics of Landing on Earth*, Cambridge, MA: MIT Press, 2020.

Lenton, Timothy M., 'Gaia and natural selection', *Nature*, 394, 1998, pp. 439–47.

Lenton, Timothy M., and Bruno Latour, 'Gaia 2.0', *Science*, 361, 2018, pp. 1066–8.

Lovelock, James, *Gaia, a New Look at Life on Earth*, Oxford University Press, 1979.

Odling Smee, John, Kevin N. Laland and Marcus W. Feldman, *Niche Construction: The Neglected Process in Evolution*, Princeton and Oxford: Princeton University Press, 2003.

Richter, Daniel deB., and Sharon A. Billings, '"One physical system": Tansley's ecosystem as Earth's critical zone', *New Phytologist*, 206, 2015, pp. 900–12.

Tansley, Arthur, 'The use and abuse of vegetational concepts and terms', *Ecology*, 16, 1935, pp. 284–307.

Vernadskij, V., *Biosfera*, Leningrad, 1926 (English edition: *The Biosfere. Complete Annotated Edition*, New York: Nevramount Publishing Company, 1998).

Wallace, Alfred R., *The Geographical Distribution of Animals*, New York: Harper, 1876.

Watson, Andrew J., and James E. Lovelock, 'Biological homeostasis of the global environment: the parable of Daisyworld', *Tellus B*, 35, 1983, pp. 286–9.

Wulf, Andrea, *The Invention of Nature: The Adventures of Alexander von Humboldt, the Last Hero of Science*, London: John Murray Publishers, 2015.

Website on the African Great Green Wall: www.unccd.int/our-work/ggwi.

7 Winds Up High and Currents in the Deep

Broecker, Wallace S., 'The Great Ocean Conveyor', *Oceanography*, 4, 1991, pp. 79–89.

Emanuel, Kerry, *What We Know about Climate Change*, Cambridge, MA: MIT Press, 2007.

Gill, Adrian, *Atmosphere–Ocean Dynamics*, New York: Academic Press, 1982.

Honh, Donovan, *Moby-Duck: The True Story of 28,800 Bath Toys Lost at Sea and of the Beachcombers, Oceanographers, Environmentalists, and Fools, Including the Author, Who Went in Search of Them*, New York: Viking, 2011.

McWilliams, James, *Fundamentals of Geophysical Fluid Dynamics*, New York: Cambridge University Press, 2006.

NOAA, *Hurricanes*: www.noaa.gov/education/resource-collections/weather -atmosphere/hurricanes.

Pedlosky, Joseph, *Ocean Circulation Theory*, Berlin and New York: Springer, 1998.

Rahmstorf, Stefan, *The Thermohaline Ocean Circulation: A Brief Fact Sheet*, 2006: www.pik-potsdam.de/~stefan/thc_fact_sheet.html.

Schneider, Tapio, 'The general circulation of the atmosphere', *Annual Review of Earth and Planetary Sciences*, 34, 2006, pp. 655–88.

Stommel, Henry, 'The westward intensification of wind-driven ocean currents', *EOS, Transactions American Geophysical Union*, 29, 1948, pp. 202–6.

Vallis, Geoffrey K., *Atmospheric and Oceanic Fluid Dynamics*, Cambridge University Press, 2006.

8 The Big Heat

Abbot Dorian S., and Eli Tziperman, 'A high-latitude convective cloud feedback and equable climates', *Quarterly Journal of the Royal Meteorological Society*, 134, 2008, pp. 165–85.

Barron, Eric J., 'Eocene Equator-to-Pole surface ocean temperatures: a significant climate problem?' *Paleoceanography*, 2, 1987, pp. 729–39.

Bender, Michael, *Paleoclimate*, Princeton University Press, 2013.

DeConto, Robert M., et al., 'Past extreme warming events linked to massive carbon release from thawing permafrost', *Nature*, 484, 2012, pp. 87–91.

Dickens, Gerald R., 'Down the Rabbit Hole: toward appropriate discussion of methane release from gas hydrate systems during the Paleocene–Eocene thermal maximum and other past hyperthermal events', *Climate of the Past*, 7, 2011, pp. 831–46.

Dickens, Gerald R., James R. O'Neil, David K. Rea and Robert M. Owen, 'Dissociation of oceanic methane hydrate as a cause of the carbon isotope excursion at the end of the Paleocene', *Paleoceanography and Paleoclimatology*, 10, 1995, pp. 965–71.

Farrell, Brian F., 'Equable climate dynamics', *Journal of Atmospheric Sciences*, 47, 1990, pp. 2986–95.

Greenwood David R., and Scott L. Wing, 'Eocene continental climates and latitudinal temperature gradients', *Geology*, 23, 1995, pp. 1044–8.

Kennett, James P., and Lowell D. Stott, 'Abrupt deep-sea warming, palae-oceanographic changes and benthic extinctions at the end of the Palaeocene', *Nature*, 353, 1991, pp. 225–9.

Korty, Robert L., Kerry A. Emanuel and Jeffery R. Scott, 'Tropical cyclone-induced upper-ocean mixing and climate: application to equable climates', *Journal of Climate*, 21, 2008, pp. 638–54.

Piana, Mark E., *Equable Climate Dynamics*: www.seas.harvard.edu/climate/eli/research/equable/index.html.

Sloan, Lisa C., and David Pollard, 'Polar stratospheric clouds: a high latitude warming mechanism in an ancient greenhouse world', *Geophysical Research Letters*, 25, 1998, pp. 3517–20.

Upchurch, Garland, et al., 'Vegetation–atmosphere interactions and their role in global warming during the latest Cretaceous', *Philosophical Transactions of The Royal Society B: Biological Sciences*, 353, 1998, pp. 97–112.

Winguth, Arne M. R., 'The Paleocene–Eocene Thermal Maximum: feedbacks between climate change and biogeochemical cycles', in Juan Blanco (ed.), *Climate Change: Geophysical Foundations and Ecological Effects*, London: IntechOpen, 2011: www.intechopen.com/books/climate-change-geophysical-foundations-and-ecological-effects/the-paleocene-eocene-thermal-maximum-feedbacks-between-climate-change-and-biogeochemical-cycles.

Zachos, James, Mark Pagani, Lisa Sloan, Ellen Thomas, and Katharina Billups, 'Trends, rhythms, and aberrations in global climate 65 Ma to present', *Science*, 292, 2001, pp. 686–93.

9 Rain, Snow and Clouds: The Planetary Water Cycle

Battisti, David S., Daniel J. Vimont and Benjamin P. Kirtman, '100 years of progress in understanding the dynamics of coupled atmosphere–ocean variability', *Meteorological Monographs*, 59, 2019.

Bechtold, Peter, *Atmospheric Thermodynamics*, Reading: ECMWF, 2015: www.ecmwf.int/sites/default/files/elibrary/2015/16954-atmospheric -thermodynamics.pdf.

Bjerknes, Jacob, 'A possible response of the atmospheric Hadley circulation to equatorial anomalies of ocean temperature', *Tellus*, 18, 1966, pp. 820–9.

Blanchard, Duncan, *From Raindrops to Volcanoes*, Mineola, NY: Dover, 1967.

Dessler, Andrew E., Zhibo Zhang and Phoebe Yang, 'Water-vapor climate feedback inferred from climate fluctuations, 2003–2008', *Geophysical Research Letters*, 35, L20704, 2008.

Fermi, Enrico, *Thermodynamics*, New York: Prentice-Hall, 1937.

Harvey, Chelsea, 'Clouds may hold the key to future warming', *E&E News*, 27 February 2019: www.scientificamerican.com/article/clouds-may-hold -the-key-to-future-warming.

Miglietta, Mario M., 'Mediterranean tropical-like cyclones (Medicanes)', *Atmosphere*, 10, 2019, p. 206.

Philander, George S., *Our Affair with El Niño*, Princeton University Press, 2004.

Schneider, Tapio, Colleen M. Kaul and Kyle G. Pressel, 'Possible climate transitions from breakup of stratocumulus decks under greenhouse warming', *Nature Geoscience*, 12, 2019, pp. 163–7.

Steinbeck, John, *To a God Unknown*, New York: Robert O. Ballou, 1933.

Strangeways, Ian, *Precipitation: Theory, Measurement and Distribution*, Cambridge University Press, 2006.

Walker, Gilbert T., 'World weather', *Monthly Weather Review*, 56, 1928, pp. 167–70.

World Meteorological Organization, *International Cloud Atlas*, 2017: https:// cloudatlas.wmo.int/en/home.html.

10 The Planet Cools

Cerling, Thure E., Yang Wang and Jay Quade, 'Expansion of C4 ecosystems as an indicator of global ecological change in the late Miocene', *Nature*, 361, 1993, pp. 344–5.

Cerling, Thure E., et al., 'Global vegetation change through the Miocene/Pliocene boundary', *Nature*, 389, 1997, pp. 153–8.

Hsü, Kenneth J., et al., 'History of the Mediterranean salinity crisis', *Nature*, 267, 1977, pp. 399–403.

Mudelsee, Manfred, Torsten Bickert, Caroline H. Lear and Gerrit Lohmann, 'Cenozoic climate changes: a review based on time series analysis of marine benthic $\delta^{18}O$ records', *Reviews of Geophysics*, 52, 2014, pp. 333–74.

Raymo, Maureen E., and William F. Ruddiman, 'Tectonic forcing of late Cenozoic climate', *Nature*, 359, 1992, pp. 117–22.

Raymo, Maureen E., William F. Ruddiman, and Philip M. Froelich, 'Influence of late Cenozoic mountain building on ocean geochemical cycles', *Geology*, 16, 1988, pp. 649–53.

Sternai, Pietro, et al., 'Magmatic forcing of Cenozoic climate?' *Journal of Geophysical Research: Solid Earth*, 125, 2020, e2018JB016460.

11 The Breath of the Ice

Adhémar, Joseph A., *Révolutions de la mer*, privately published in France, 1842.

Agassiz, Louis, *Etudes sur les glaciers*, Neuchâtel: Jent et Gassmann, 1840.

Baldovin, Marco, Fabio Cecconi, Antonello Provenzale and Angelo Vulpiani, 'Extracting causation from millenial-scale climate fluctuations in the last 800 ky', *Scientific Reports*, 12, 2022, 15320.

Bender, Michael L., *Paleoclimate*, Princeton University Press, 2013.

Benzi, Roberto, Giorgio Parisi, Alfonso Sutera and Angelo Vulpiani, 'Stochastic resonance in climatic change', *Tellus*, 34, 1982, pp. 10–16.

Bereiter, Bernhard, et al., 'Revision of the EPICA Dome C CO_2 record from 800 to 600 kyr before present', *Geophysical Research Letters*, 42, 2014, pp. 542–9.

Broecker, Wallace S., *What Drives the Glacial Cycles?* New York: Eldigio Press, 2015.

Clark, Peter U., and David Pollard, 'Origin of the Middle Pleistocene Transition by ice sheet erosion of regolith', *Paleoceanography and Paleoclimatology*, 13, 1998, pp. 1–9.

Croll, James, 'On the physical cause of the change of climate during geological epochs', *Philosophical Magazine*, 28, 1864, pp. 121–37.

Fagan, Brian, *Cro-Magnon: How the Ice Age Gave Birth to the First Modern Humans*, London: Bloomsbury Press, 2013.

Fuller-Wright, Liz, *What Caused the Ice Ages? Tiny Ocean Fossils Offer Key Evidence*, 10 December 2020: www.princeton.edu/news/2020/12/10/what-caused-ice-ages-tiny-ocean-fossils-offer-key-evidence.

Ghil, Michael, and Hervé Le Treut, 'A climate model with cryodynamics and geodynamics', *Journal of Geophysical Research*, 86, 1981, pp. 5262–70.

Hays, J. D., John Imbrie and N. J. Shackleton, 'Variations in the Earth's orbit: pacemaker of the ice ages', *Science*, 194, 1976, pp. 1121–32.

Huybers, Peter, 'Pleistocene glacial variability as a chaotic response to obliquity forcing', *Climate of the Past*, 5, 2009, pp. 481–8.

Imbrie, John, 'Astronomical theory of the Pleistocene ice ages: a brief historical review', *Icarus*, 50, 1982, pp. 408–22.

Imbrie, John, and Katherine P. Imbrie, *Ice Ages: Solving the Mystery*, Short Hills, NJ: Enslow Publishers, 1979.

Laskar, Jacques, 'A numerical experiment on the chaotic behaviour of the Solar System', *Nature*, 338, 1989, pp. 237–8.

Lisiecki, Lorraine E., and Maureen E. Raymo, 'A Pliocene–Pleistocene stack of 57 globally distributed benthic δ18O records', *Paleoceanography and Paleoclimatology*, 20, 2005, PA1003.

Lüthi, Dieter, et al., 'High-resolution carbon dioxide concentration record 650,000–800,000 years before present', *Nature*, 453, 2008, pp. 379–82.

Milanković, Milutin, *Kanon der Erdbestrahlung und seine Anwendung auf das Eiszeitenproblem*, Royal Serbian Academy Special Publications, Belgrade: Royal Serbian Academy132, 1941.

Milanković, Milutin, *Théorie mathématique des phénomènes thermiques produits par la radiation solaire*, Paris: Gauthier-Villars, 1920.

Nicolis, Catherine, 'Stochastic aspects of climatic transitions – response to a periodic forcing', *Tellus*, 34, 1982, pp. 1–9.

Parrenin, Frederic, et al., 'Synchronous change of atmospheric CO_2 and Antarctic temperature during the last deglacial warming', *Science*, 339, 2013, pp. 1060–3.

Poincaré, Henri, *Les méthodes nouvelles de la mécanique céleste*, 3 vols., Paris: Gauthier-Villars, 1892–9.

Raymo, Maureen E., and Peter Huybers, 'Unlocking the mysteries of the ice ages', *Nature*, 451, 2008, pp. 284–5.

Raymo, Maureen E., Lorraine E. Lisieki and Kerim N. Nisancioglu, 'Plio-Pleistocene ice volume, antarctic climate, and the global δ¹⁸O record', *Science*, 313, 2006, pp. 492–5.

Steffen, Will, et al., *Global Change and the Earth System: A Planet under Pressure*, Berlin: Springer, 2006.

Tzedakis, Chronis, Michel Crucifix, Takahito Mitsui and Eric W. Wolff, 'A simple rule to determine which insolation cycles lead to interglacials', *Nature*, 542, 2017, pp. 427–32.

Venetz, Ignaz, *Mémoire sur les variations de la température dans les Alpes de la Suisse*, 1833.

12 Agitated Ice

Alley, Richard B., 'Ice-core evidence of abrupt climate changes', *Proceedings of the National Academy of Sciences*, 97, 2000, pp. 1331–4.

Blunier, Thomas, and Edward J. Brook, 'Timing of millennial-scale climate change in Antarctica and Greenland during the last glacial period', *Science*, 291, 2001, pp. 109–12.

Bond, Gerard C., et al., 'Evidence for massive discharges of icebergs into the North Atlantic Ocean during the last glacial period', *Nature*, 360, 1992, pp. 245–9.

Broecker, Wallace S., 'Paleocean circulation during the last deglaciation: a bipolar seesaw?' *Paleoceanography*, 13, 1998, pp. 119–21.

Broecker, Wallace S., 'Was the Younger Dryas triggered by a flood?' *Science*, 312, 2006, pp. 1146–8.

Clark, Peter U., et al., 'Global climate evolution during the last deglaciation', *Proceedings of the National Academy of Sciences*, https://doi.org/10.1073/pnas.1116619109.

Croll, James, 'On ocean currents', *London, Edinburgh, and Dublin Philosophical Magazine and Journal of Science*, 39, 1870, pp. 81–106.

Crowley, Thomas J., 'North Atlantic deep water cools the Southern hemisphere', *Paleoceanography and Paleoclimatology*, 7, 1992, pp. 489–97.

Dansgaard, Willi, et al., 'Evidence for general instability of past climate from a 250-kyr ice-core record', *Nature*, 364, 1993, pp. 218–20.

Greenland Ice-core Project (GRIP) Members, 'Climate instability during the last interglacial period recorded in the GRIP ice core', *Nature*, 364, 1993, pp. 203–7.

Grootes, Pieter M., Minze Stuiver, James W. C. White, Sigfús Johnsen and Jean Jouzel, 'Comparison of oxygen isotope records from the GISP2 and GRIP Greenland ice cores', *Nature*, 366, 1993, pp. 552–4.

Heinrich, Hartmut, 'Origin and consequences of cyclic ice rafting in the Northeast Atlantic Ocean during the past 130,000 years', *Quaternary Research*, 29, 1988, pp. 142–52.

Lambeck, Kurt, et al., 'Sea level and global ice volumes from the Last Glacial Maximum to the Holocene', *Proceedings of the National Academy of Sciences*, 111, 2014, pp. 15296–303.

Menviel, Laurie C., Luke C. Skinner, Lev Tarasov and Polychronis C. Tzedakis, 'An ice–climate oscillatory framework for Dansgaard–Oeschger cycles', *Nature Reviews Earth & Environment*, 1, 2020, pp. 677–93.

Schulz, Michael, 'On the 1470-year pacing of Dansgaard–Oeschger warm events', *Paleoceanography and Paleoclimatology*, 17, 2002, pp. 41–9.

Shakun, Jeremy D., et al., 'Global warming preceded by increasing carbon dioxide concentrations during the last deglaciation', *Nature*, 484, 2012, pp. 49–55.

Stocker, Thomas F., Daniel G. Wright and Lawrence A. Mysak, 'A zonally averaged, coupled ocean–atmosphere model for paleoclimate studies', *Journal of Climate*, 5, 1992, pp. 773–97.

Stocker, Thomas F., and Sigfús J. Johnsen, 'A minimum thermodynamic model for the bipolar seesaw', *Paleoceanography and Paleoclimatology*, 18, 2003.

von Storch, Hans, and Kay C. Emeis, *Hartmut Heinrich – the Unknown World Famous Climate Researcher of Hamburg*, 2017: www.researchgate .net/publication/316171949_Hartmut_Heinrich_-_the_unknown_world _famous_climate_researcher_of_Hamburg.

13 Conquering the Planet

Bartlein, P. J., et al., 'Pollen-based continental climate reconstructions at 6 and 21 ka: a global synthesis', *Climate Dynamics*, 2010, pp. 775–802.

Blom, Philipp, *Nature's Mutiny: How the Little Ice Age of the Long Seventeenth Century Transformed the West and Shaped the Present*, New York: Liveright Publishing Corporation, 2019.

Büntgen, U., and W. Tegel, 'European tree-ring data and the Medieval Climate Anomaly', *Past Global Changes Magazine*, 19, 1, 2011, pp. 14–15.

Camenisch, Chantal, et al., 'Climate reconstruction and impacts from the archives of societies', *Past Global Changes Magazine*, 28, 2, 2020, pp. 33–68.

Camuffo, Dario, et al., 'When the Lagoon was frozen over in Venice from A.D. 604 to 2012: evidence from written documentary sources, visual arts and instrumental readings', *Méditerranée*, 2017: http://mediterranee.revues.org/7983.

Diamond, Jared, *Collapse: How Societies Choose to Fail or Succeed*, New York: Viking, 2005.

Diamond, Jared, *Guns, Germs, and Steel. The Fates of Human Societies*, New York: W. W. Norton, 1997.

Fagan, Brian, *The Little Ice Age*, New York: Basic Books, 2000.

Fritts, Harold C., *Tree Rings and Climate*, Cambridge, MA: Academic Press, 1976.

Graeber, David, and David Wendrow, *The Dawn of Everything*, New York: Farrar, Straus and Giroux, 2021.

Kaufman, Darrell, et al., 'A global database of Holocene paleotemperature records', *Scientific Data*, 7, 2020, p. 115.

Kaufman, Darrell, et al., 'Holocene global mean surface temperature, a multi-method reconstruction approach', *Scientific Data*, 7, 2020, p. 201.

Lamb, Hubert H., *Climate, History and the Modern World*, London: Routledge, 1982.

Le Roy Ladurie, Emmanuel, *Histoire du climat depuis l'an mil* [*History of the Climate since the Year 1000*], Paris: Flammarion, 1967

Marcott, Shaun A., et al., 'A reconstruction of regional and global temperature for the past 11,300 years', *Science*, 339, 2013, p. 1198.

NOAA: www.ncdc.noaa.gov/global-warming/mid-holocene-warm-period.

Owens, Mathew J., et al., 'The Maunder minimum and the Little Ice Age: an update from recent reconstructions and climate simulations', *Journal of Space Weather and Space Climate*, 7, A33, 2017.

PAGES 2k Consortium, 'Continental-scale temperature variability during the past two millennia', *Nature Geoscience*, 6, 2013, pp. 339–46.

PAGES 2k Consortium, 'A global multiproxy database for temperature reconstructions of the Common Era', *Scientific Data*, 4, 2017, 170088.

Park Williams, A., et al., 'Large contribution from anthropogenic warming to an emerging North American megadrought', *Science*, 368, 2020, pp. 314–18.

Tierney, Jessica E., Francesco S. R. Pausata and Peter B. deMenocal, 'Rainfall regimes of the Green Sahara', *Science Advances*, 3, 2017, e1601503.

14 The Age of Humanity
American Meteorological Society (AMS), *State of the Climate*, 2019: www.ametsoc.org/index.cfm/ams/publications/bulletin-of-the-american-meteorological-society-bams/state-of-the-climate.

Boden, T. A., G. Marland and R. J. Andres, *Global, Regional, and National Fossil-Fuel CO_2 Emissions*, Oak Ridge, TN: Carbon Dioxide Information Analysis Center, Oak Ridge National Laboratory, US Department of Energy, 2017.

Crutzen, Paul J., and Eugene F. Stoermer, 'The "Anthropocene"', *IGBP Newsletter*, 41, 2000, pp. 17–18.

European Environment Agency, *Europe's State of the Environment 2020*, 2020: www.eea.europa.eu/highlights/soer2020-europes-environment-state-and-outlook-report.

Friedlingstein, Pierre, et al., 'Global Carbon Budget 2019', *Earth System Science Data*, 11, 2019, pp. 1783–1838.

Intergovernmental Panel on Climate Change (IPCC), *Sixth Assessment Report (AR6)*, 2021: www.ipcc.ch/assessment-report/ar6.

Keeling, Charles D., 'The concentration and isotopic abundances of carbon dioxide in the atmosphere', *Tellus*, 12, 1960, pp. 200–3.

Keeling, Charles D., et al., *Exchanges of Atmospheric CO_2 and $13CO_2$ with the Terrestrial Biosphere and Oceans from 1978 to 2000. I: Global Aspects*, University of California San Diego: Library – Scripps Digital Collection, 2001: https://escholarship.org/uc/item/09v319r9.

Kolbert, Elizabeth, *The Sixth Extinction: An Unnatural History*, New York: Henry Holt and Company, 2014.

Lüthi, Dieter, et al., 'High-resolution carbon dioxide concentration record 650,000–800,000 years before present', *Nature*, 453, 2008, pp. 379–82

MacFarling Meure, C., et al., 'Law Dome CO_2, CH4 and N2O ice core records extended to 2000 years BP', *Geophysical Research Letters*, 33, 14, 2006.

Meadows, Donella H., Dennis L. Meadows, Jorgen Randers and William W. Behrens III, *The Limits to Growth*, Falls Church: Potomac Associates, 1972.

Orcutt, Beth N., Isabelle Daniel and Rajdeep Dasgupta (eds.), *Deep Carbon: Past to Present*, Cambridge University Press, 2020.

Ruddiman, William, *Earth's Climate: Past and Future*, New York: W. H. Freeman & Co., 2001.

15 Global Warming

Bishop, D. A., A. Park Williams and R. Seager, 'Increased Fall precipitation in the southeastern United States driven by higher-intensity, frontal precipitation', *Geophysical Research Letters*, 46, 2019, pp. 8300–9.

Copernicus Atmosphere Monitoring Service (CAMS), *Wildfires Wreaked Havoc in 2021, CAMS Tracked Their Impact*: https://atmosphere.copernicus .eu/wildfires-wreaked-havoc-2021-cams-tracked-their-impact.

Eicker, Annette, et al., 'Does GRACE see the terrestrial water cycle "intensifying"?' *Journal of Geophysical Research – Atmospheres*, 121, 2016, pp. 733–45.

European Environmental Agency (EEA), *Indicator Assessment: Mean Precipitation*, 2017: www.eea.europa.eu/data-and-maps/indicators /european-precipitation-2/assessment.

European Environmental Agency (EEA), *Trends in Summer Soil Moisture in Europe*, 2017: www.eea.europa.eu/data-and-maps/figures/trends-in -summer-soil-moisture-1.

Giorgi, Filippo, 'Climate change hot-spots', *Geophysical Research Letters*, 33, 2006, L08707.

Giorgi, Filippo, et al., 'Higher hydroclimatic intensity with global warming', *Journal of Climate*, 20, 2011, pp. 5309–24.

Held, Isaac M., 'The cause of the pause', *Nature*, 501, 2013, pp. 318–19.

Huntington, Thomas G., Peter K. Weiskel, David M. Wolock and Gregory J. McCabe, 'A new indicator framework for quantifying the intensity of the terrestrial water cycle', *Journal of Hydrology*, 559, 2018, pp. 361–72.

Intergovernmental Panel on Climate Change (IPCC), *Global Warming of 1.5 °C*, 2019: www.ipcc.ch/sr15.

Intergovernmental Panel on Climate Change (IPCC), *Sixth Assessment Report (AR6)*, 2021: www.ipcc.ch/assessment-report/ar6.

Kosaka, Yu, and Shang-Ping Xie, 'Recent global-warming hiatus tied to equatorial Pacific surface cooling', *Nature*, 501, 2013, pp. 403–7.

NASA, *Drought Makes Its Home on the Range*: https://climate.nasa.gov/news /3117.

NASA, *NASA-Led Study Reveals the Causes of Sea Level Rise since 1900*: https://climate.nasa.gov/news/3012/nasa-led-study-reveals-the-causes-of -sea-level-rise-since-1900.

Turco, Marco, Elisa Palazzi, Jost von Hardenberg and Antonello Provenzale, 'Observed climate change hotspots', *Geophysical Research Letters*, 42, 2015.

Turco, Marco, et al., 'Decreasing fires in Mediterranean Europe', *PLoS ONE*, 11, 2016, e0150663.

Turco, Marco, et al., 'On the key role of droughts in the dynamics of summer fires in Mediterranean Europe', *Scientific Reports*, 7, 2017, p. 81.

US Environmental Protection Agency (EPA), *Climate Change Indicators: U.S. and Global Precipitation*: www.epa.gov/climate-indicators/climate -change-indicators-us-and-global-precipitation.

16 Arctic Sentinels

'Arctic change and mid-latitude weather', *Nature*, 2019: www.nature.com /collections/bfihgidbhc.

Biskaborn, Boris K., et al., 'Permafrost is warming at a global scale', *Nature Communications*, 10, 2019, p. 264.

Boike, Julia, et al., 'A 20-year record (1998–2017) of permafrost, active layer and meteorological conditions at a high Arctic permafrost research site (Bayelva, Spitsbergen)', *Earth System Science Data*, 10, 2018, pp. 355–90.

Caesar, Levke, et al., 'Current Atlantic Meridional Overturning Circulation weakest in last millennium', *Nature Geoscience*, 14, 2021, pp. 118–20.

Conservation of Arctic Flora and Fauna, CAFF, *Arctic Biodiversity Assessment*, 2013: www.caff.is and www.arcticbiodiversity.is.

GRID Arendal, *Distribution and Current Trend of Polar Bear Subpopulations throughout the Circumpolar Arctic*, 2010: www.grida.no/resources/7757.

The IMBIE Team, 'Mass balance of the Greenland ice sheet from 1992 to 2018', *Nature*, 579, 2020, pp. 233–9.

Magnani, Marta, et al., 'Microscale drivers of summer CO_2 fluxes in the Svalbard High Arctic tundra', *Scientific Reports*, 12, 2022, p. 763: https:// doi.org/10.1038/s41598-021-04728-0.

Myers-Smith, Isla, et al., 'Complexity revealed in the greening of the Arctic', *Nature Climate Change*, 10, 2020, pp. 106–17.

Parkinson, Claire, 'A 40-y record reveals gradual Antarctic sea ice increases followed by decreases at rates far exceeding the rates seen in the Arctic', *Proceedings of the National Academy of Sciences*, 116, 2019, pp. 14414–23.

Parmesan, Camille, 'Ecological and evolutionary responses to recent climate change', *Annual Review of Ecology, Evolution and Systematics*, 37, 2006, pp. 637–69.

Pedersen, Åshild, et al., 'Five decades of terrestrial and freshwater research at Ny-Ålesund, Svalbard: current status and knowledge gaps', *Polar Research*, 41, 2022, 6310.

Post, Eric, et al., 'Ecological dynamics across the Arctic associated with recent climate change', *Science*, 325, 2009, p. 1355.

Prop, Jouke, et al., 'Climate change and the increasing impact of polar bears on bird populations', *Frontiers in Ecology and Evolution*, 25 March 2015: www.frontiersin.org/articles/10.3389/fevo.2015.00033/full.

Saino, Nicola, et al., 'Climate warming, ecological mismatch at arrival and population decline in migratory birds', *Proceedings of the Royal Society B: Biological Sciences*, 278, 2010, pp. 835–42.

Schuur, Ted, *Permafrost and the Global Carbon Cycle*, NOAA's Arctic Program: https://arctic.noaa.gov/Report-Card/Report-Card-2019/ArtMID/7916 /ArticleID/844/Permafrost-and-the-Global-Carbon-Cycle.

Seager, Richard, et al., 'Is the Gulf Stream responsible for Europe's mild winters?' *Quarterly Journal of the Royal Meteorological Society*, 128, 2002, pp. 2563–86.

Serreze, Mark C., and Roger G. Barry, 'Processes and impacts of Arctic amplification: a research synthesis', *Global and Planetary Change*, 77, 2011, pp. 85–96.

Steiner, Nadja S., et al., 'Impacts of the changing ocean–sea ice system on the key forage fish Arctic Cod (*Boreogadus saida*) and subsistence fisheries in the Western Canadian Arctic – evaluating linked climate, ecosystem and economic (CEE) models', *Frontiers in Marine Science*, 10 April 2019: frontiersin.org/articles/10.3389/fmars.2019.00179/full.

Straneo, Fiamma, and Claudia Cenedese, 'The dynamics of Greenland's glacial fjords and their role in climate', *Annual Review of Marine Science*, 7, 2015, pp. 89–112.

Tesi, Tommaso, et al., 'Massive remobilization of permafrost carbon during post-glacial warming', *Nature Communications*, 7, 2016, p. 13653.

Wadhams, Peter, *A Farewell to Ice*, London: Penguin, 2016.

17 *The Mountain Heat*

Beckage, Brian, et al., 'A rapid upward shift of a forest ecotone during 40 years of warming in the Green Mountains of Vermont', *Proceedings of the National Academy of Sciences*, 105, 2008, pp. 4197–4202.

Beniston, Martin, 'Climatic change in mountain regions: a review of possible impacts', *Climatic Change*, 59, 2003, pp. 5–31.

Beniston, Martin, et al., 'The European mountain cryosphere: a review of its current state, trends, and future challenges', *The Cryosphere*, 12, 2018, pp. 759–94.

Cazzolla Gatti, Roberto, et al., 'Accelerating upward treeline shift in the Altai Mountains under last-century climate change', *Scientific Reports*, 9, 2019, p. 7678.

Cordero, Raul R., et al., 'Dry-season snow cover losses in the Andes (18°–40°S) driven by changes in large-scale climate modes', *Scientific Reports*, 9, 2019, p. 16945.

Diaz, Henry F., and Raymond S. Bradley, 'Temperature variations during the last century at high elevation sites', *Climatic Change*, 36, 1997, pp. 253–79.

Ehrlich, Daniele, Michele Melchiorri, and Claudia Capitani, 'Population trends and urbanisation in mountain ranges of the world', *Land*, 10, 2021, p. 255: https://doi.org/10.3390/land10030255.

European Environmental Agency, *Regional Climate Change and Adaptation: The Alps Facing the Challenge of Changing Water Resources*, EEA Report 8/2009: www.eea.europa.eu/publications/alps-climate-change-and-adaptation-2009.

Farinotti, Daniel, et al., 'Manifestations and mechanisms of the Karakoram glacier anomaly', *Nature Geoscience*, 13, 2020, pp. 8–16.

Fassnacht, Steven R., et al., 'Sub-seasonal snowpack trends in the Rocky Mountain National Park Area, Colorado, USA', *Water*, 10, 2018, p. 562.

Filippi, Luca, Elisa Palazzi, Jost von Hardenberg and Antonello Provenzale, 'Multidecadal variations in the relationship between the NAO and winter precipitation in the Hindu Kush – Karakoram', *Journal of Climate*, 27, 2014, pp. 7890–902.

Freeman, Benjamin G., et al., 'Climate change causes upslope shifts and mountaintop extirpations in a tropical bird community', *Proceedings of the National Academy of Sciences*, 115, 2018, pp. 11982–7.

Gardner, A.S., et al., 'A reconciled estimate of glacier contributions to sea level rise: 2003 to 2009', *Science*, 340, 2013, pp. 852–7: doi:10.1126/science.1234532.

Immerzeel, W. W., et al., 'Importance and vulnerability of the world's water towers', *Nature*, 577, 2020, pp. 364–9.

Intergovernmental Panel on Climate Change, *Special Report on the Ocean and the Cryosphere in a Changing Climate*, 2019: www.ipcc.ch/srocc.

Jacobson, Andrew R., et al., 'Climate forcing and density dependence in a mountain ungulate population', *Ecology*, 85, 2005, pp. 1598–610.

Jin, Qinjian, et al., 'Interactions of Asian mineral dust with Indian summer monsoon: recent advances and challenges', *Earth-Science Reviews*, 215, 2021, p. 103562.

Kammer, Peter M., Christian Schöb and Philippe Choler, 'Increasing species richness on mountain summits: upward migration due to anthropogenic climate change or re-colonisation?' *Journal of Vegetation Science*, 18, 2007, pp. 301–6.

Katzenberger, A., J. Schewe, J. Pongratz, A. Levermann, 'Robust increase of Indian monsoon rainfall and its variability under future warming in CMIP6 models', *Earth System Dynamics*, 12, 2021, pp. 367–86.

Magnani, Marta, et al., 'Drivers of carbon fluxes in Alpine tundra: a comparison of three empirical model approaches', *Science of the Total Environment*, 732, 2020, p. 139139.

McCain, Christy M., Sarah R. B. King, and Tim M. Szewczyk, 'Unusually large upward shifts in cold-adapted, montane mammals as temperature warms', *Ecology*, 102, 2021, e03300.

Minder, J. R., T. W. Letcher and C. Liu, 'The character and causes of elevation-dependent warming in high-resolution simulations of Rocky Mountain climate change', *Journal of Climate*, 31, 2018, pp. 2093–113.

Palazzi, Elisa, Luca Mortarini, Silvia Terzago and Jost von Hardenberg, 'Elevation-dependent warming in global climate model simulations at high spatial resolution', *Climate Dynamics*, 52, 2019, pp. 2685–2702.

Parmesan, Camille, 'Ecological and evolutionary responses to recent climate change', *Annual Review of Ecology, Evolution and Systematics*, 37, 2006, pp. 637–69.

Pepin, Nicholas, et al., 'Climate Changes and their elevational patterns in the mountains of the world', *Reviews of Geophysics*, 60, 2022, e2020RG000730.

Pepin, Nicholas, et al. (Mountain Research Initiative EDW Working Group), 'Elevation-dependent warming in mountain regions of the world', *Nature Climate Change*, 5, 2015, pp. 424–30.

Pettorelli, Nathalie, et al., 'Early onset of vegetation growth vs. rapid green-up: impacts on juvenile mountain ungulates', *Ecology*, 88, 2007, pp. 381–90.

Rangwala, Imtiaz, 'Amplified water vapour feedback at high altitudes during winter', *International Journal of Climatology*, 2012: https://doi.org/10.1002/joc.3477.

Rangwala, I., and J. R. Miller, 'Climate change in mountains: a review of elevation-dependent warming and its possible causes', *Climatic Change*, 114, 2012, pp. 527–47.

Sabin, T. P., et al., 'Climate change over the Himalayas', in *Assessment of Climate Change over the Indian Region*, Singapore: Springer, 2020, pp. 207–22.

Scherler, Dirk, Bodo Bookhagen and Manfred R. Strecker, 'Spatially variable response of Himalayan glaciers to climate change affected by debris cover', *Nature Geoscience*, 2011: https://doi.org/10.1038/NGEO1068.

Sritharan, Meena S., Frank A. Hemmings and Angela T. Moles, 'Few changes in native Australian alpine plant morphology, despite substantial local climate change', *Ecology and Evolution*, 11, 2021, pp. 4854–65.

Stefaniak, Anne, et al., 'Mass balance and surface evolution of the debris-covered Miage Glacier, 1990–2018', *Geomorphology*, 373, 2021, 107474.

Toledo, Osmar, Elisa Palazzi, Ivan Mauricio Cely Toro and Luca Mortarini, 'Comparison of elevation-dependent warming and its drivers in the tropical and subtropical Andes', *Climate Dynamics*, 2021: https://doi.org /10.1007/s00382-021-06081-4.

United Nations, *Sustainable Mountain Development*, Resolution A/ Res/62/196, 2008.

Viterbi, Ramona, Cristiana Cerrato, Radames Bionda and Antonello Provenzale, 'Effects of temperature rise on multi-taxa distributions in mountain ecosystems', *Diversity*, 12, 2020, p. 210.

Viviroli, D., et al., 'Climate change and mountain water resources: overview and recommendations for research, management and policy', *Hydrology and Earth System Sciences*, 15, 2011, pp. 471–504.

Walther, Gian-Reto, Sascha Beißner and Conradin A.Burga, 'Trends in the upward shift of Alpine plants', *Journal of Vegetation Science*, 16, 2005, pp. 541–8.

Wang, Pin Xian, et al., 'The global monsoon across time scales: mechanisms and outstanding issues', *Earth-Science Reviews*, 174, 2017, pp. 84–121.

Zemp, M., et al., 'Global glacier mass changes and their contributions to sea-level rise from 1961 to 2016', *Nature*, 568, 2019, pp. 382–6.

Zu, Kuiling, et al., 'Upward shift and elevational range contractions of subtropical mountain plants in response to climate change', *Science of the Total Environment*, 783, 2021, 146896.

18 Digital Twins

Bauer, Peter, Bjorn Stevens and Wilco Hazeleger, 'A digital twin of Earth for the green transition', *Nature Climate Change*, 11, 2021, pp. 80–3.

Chibbaro, Sergio, Lamberto Rondoni and Angelo Vulpiani, *Reductionism, Emergence and Levels of Reality*, New York: Springer, 2014.

Cowtan, Kevin, et al., 'Robust comparison of climate models with observations using blended land air and ocean sea surface temperatures', *Geophysical Research Letters*, 42, 2015, pp. 6526–34.

European Commission, *Destination Earth (DestinE)*, 2020: https://ec.europa .eu/digital-single-market/en/destination-earth-destine.

European Space Agency, *Digital Twin Earth*, 2020: www.esa.int/Applications /Observing_the_Earth/Working_towards_a_Digital_Twin_of_Earth.

Ghil, Michael, and V. Lucarini, 'The physics of climate variability and climate change', *Reviews of Modern Physics*, 92, 2020, 035002.

Giorgi, Filippo, 'Thirty years of regional climate modeling: where are we and where are we going next?', *Journal of Geophysical Research – Atmospheres*, 124, 2019, pp. 5696–723.

Hausfather, Zeke, et al., 'Evaluating the performance of past climate model projections', *Geophysical Research Letters*, 47, 2019, e2019GL085378.

Hofstadter, Douglas R., *Gödel, Escher, Bach: An Eternal Golden Braid*, New York: Basic Books, 1979.

Lloyd, Elisabeth, and Eric Winsberg (eds.), *Climate Modelling: Philosophical and Conceptual Issues*, London: Palgrave Macmillan, 2018.

Lorenz, Edward N., 'Deterministic nonperiodic flow', *Journal of the Atmospheric Sciences*, 20, 1963, pp. 130–41.

McGuffie, Kendal, and Ann Henderson-Sellers, *The Climate Modelling Primer*, Hoboken, NJL Wiley, 2014.

Meehl, Gerald D., et al., 'Combinations of natural and anthropogenic forcings in twentieth-century climate', *Journal of Climate*, 17, 2004, pp. 3721–7.

Pierrehumbert, Raymond T., *Principles of Planetary Climates*, Cambridge University Press, 2010.

Provenzale, Antonello, 'Climate models', *Rendiconti Lincei. Scienze Fisiche e Naturali*, 25, 2014, pp. 49–58.

Provenzale, Antonello, Elisa Palazzi and Klaus Fraedrich (eds.), *The Fluid Dynamics of Climate*, Berlin: Springer, 2016.

Schmidt, Gavin, *Climate Model Projections Compared to Observations*, RealClimate: Climate Science from Climate Scientists, 2021: www

.realclimate.org/index.php/climate-model-projections-compared-to
-observations/#ITEM-20317-4.

Schmidt, Gavin, Drew T. Shindell and Kostas Tsigaridis, 'Reconciling warming trends', *Nature Geoscience*, 7, 2014, pp. 158–60.

Schneider, Tapio, et al., 'Climate goals and computing the future of clouds', *Nature Climate Change*, 7, 2017, pp. 3–5.

Smith, Leonard A., *Chaos: A Very Short Introduction*, Oxford University Press, 2007.

Trenberth, Kevin (ed.), *Climate System Modeling*, Cambridge University Press, 1992.

WCRP, World Climate Research Programme, *Coupled Model Intercomparison Project*: www.wcrp-climate.org/wgcm-cmip.

19 Knowing in Order to Anticipate, Anticipating in Order to Act

Bonanno, Riccardo, Christian Ronchi, Barbara Cagnazzi and Antonello Provenzale, 'Glacier response to current climate change and future scenarios in the northwestern Italian Alps', *Regional Environmental Change*, 14, 2013, pp. 633–43.

Bonatti, Enrico, 'Darwin and inequality', *Substantia*, 5, 1, 2021: https://doi.org/10.36253/Substantia-1121.

Boschi, Chiara, et al., 'Brucite-driven CO_2 uptake in serpentinized dunites (Ligurian Ophiolites, Montecastelli, Tuscany)', *Lithos*, 2017, pp. 288–9, 264–81.

CarbFix, www.carbfix.com.

Ghil, M., and S. Childress, *Topics in Geophysical Fluid Dynamics: Atmospheric Dynamics, Dynamo Theory and Climate Dynamics*, Berlin: Springer Science & Business Media, 1987: https://doi.org/10.1007/978-1-4612-1052 -8, ISBN 978-0-387-96475-l, reissued as an e-book in 2012.

Intergovernmental Panel on Climate Change (IPCC), *Global Warming of 1.5 °C*, 2018: www.ipcc.ch/sr15.

Intergovernmental Panel on Climate Change (IPCC), *Sixth Assessment Report (AR6)*, 2021: www.ipcc.ch/assessment-report/ar6.

Lenton, T. M., et al., 'Tipping elements in the Earth's climate system', *Proceedings of the National Academy of Sciences*, 105, 2008, pp. 1786–93.

Lenton, T. M., et al., 'Climate tipping points: too risky to bet against', *Nature*, 575, 2019, pp. 592–5.

Lorenz, Edward N., 'Climatic change as a mathematical problem', *Journal of Applied Meteorology and Climatology*, 9, 1970, pp. 325–9.

Lynch, Peter, *The Emergence of Numerical Weather Prediction*, Cambridge University Press, 2006.

Moss, Richard, et al., 'The next generation of scenarios for climate change research and assessment', *Nature*, 463, 2010, pp. 747–56.

Riahi, Keywan, et al., 'The Shared Socioeconomic Pathways and their energy, land use, and greenhouse gas emissions implications: an overview', *Global Environmental Change*, 42, 2017, pp. 153–68.

Richardson, Lewis F., *Weather Prediction by Numerical Process*, Cambridge University Press, 1922.

Rogelj, Joeri, et al., 'Energy system transformations for limiting end-of-century warming to below 1.5 °C', *Nature Climate Change*, 5, 2015, pp. 519–28.

Sgouridis, Sgouris, Denes Csala and Ugo Bardi, 'The sower's way: quantifying the narrowing net-energy pathways to a global energy transition', *Environmental Research Letters*, 11, 2016.

Taucher, Jan, et al., 'Changing carbon-to-nitrogen ratios of organic-matter export under ocean acidification', *Nature Climate Change*, 11, 2021, pp. 52–7.

Tilmes, Simone, et al., 'The hydrological impact of geoengineering in the Geoengineering Model Intercomparison Project (GeoMIP)', *Journal of Geophysical Research – Atmospheres*, 118, 2013, pp. 11036–58.

Turco, Marco, et al., 'Exacerbated fires in Mediterranean Europe due to anthropogenic warming projected with non-stationary climate-fire models', *Nature Communications*, 9, 2018, p. 3821.

World Bank, *Gini Index – World Bank Estimate*: https://data.worldbank.org /indicator/SI.POV.GINI?end=2019&start=2019&view=map&year=2019.

Conclusion: The Journey Continues

Attenborough, David, *A Life on Our Planet*, documentary, and book published by Ebury Publishing, London, 2020.

European Commission, Nature Based Solutions: https://ec.europa.eu /info/research-and-innovation/research-area/environment/nature-based -solutions_en.

IUCN, Nature Based Solutions: www.iucn.org/theme/nature-based-solutions.

Jensen, Cecile, *Working with Instead of against Nature Is the Only Way Out of the Climate and Biodiversity Crisis*: https://phys.org/news/2021-11-nature -climate-biodiversity-crisis.html.

Lovelock, James, *Novacene: The Coming Age of Hyperintelligence*, London: Penguin Books Limited, 2019.

Index

Abbot, Dorian 97–8
Adhémar, Joseph-Alphonse 132
aerosols
  carbonaceous or sulphur-rich 174–5
  monsoons and 218
Africa
  climate change hotspots 178
  wildfires 183
Agassiz, Louis 126
agriculture
  bad weather and 108
  deforestation and 168, 169
  estimating climate impacts 247
  nitrogen fertilizers 170
  rice paddies 168
Alaska
  Arctic research stations 202
  Bering Strait isthmus 139–40
  Eocene fossils 94–5
albedo
  Arctic region and 192, 196
  calculating effect of 37–8
  clouds 111, 113
  Daisyworld parable 75
  effect of 2, 31–2, 32–3, 138
  fires and 204
  glaciation and 25, 135
  human activity and 188
  mountains and 210
  polar Eocene plants 98
  vegetation and 69–70, 70
algae
  biological carbon cycle 165
  development of 41
  glacial conditions and 26

  symbiosis with fungi 58
Alley, Richard 142
Alps mountains
  climate studies 210
  dendroclimatic study of 154, 154
  Gran Paradiso National Park 220–3, 221
  length of glaciers 214
  Miage, black glacier 213–14, 215
  RCP glaciers scenario 248, 248
  warming in 209
Alps mountains, Australian 220
Álvarez, Luis 61–3, 62
Álvarez, Walter 61–3, 62
Amazonia
  climate change hotspots 178
  evapotranspiration 71
  predicting deforestation 243
ammonites, mass extinction of 62
Amundsen–Nobile Climate Change
    Tower 202–3, 203
Andes mountains
  birds of 220
  climate studies 210
  water storage 215
animals
  biological carbon cycle 165
  herbivores and C_4 cycle 122
  herbivores in prairies 118
  pine processionary moth 220
  polar bears 197
  species replacement 201
  see also life, early development of
Antarctica
  compared to Arctic 193
  end of Last Glacial Maximum 146

carbon (*cont.*)
  in permafrost matter 200
  RuBisCO 121–2
  soot 174–5, 183, 210
  vast coal burning 61
carbon dioxide
  Archean Eon 7, 15
  carbon capture and storage 255–6
  Cenozoic period 117, 135
  complex interactions of 130–1
  crucial role 2, 34–7, 138
  delay in temperature rise 130–1
  Eocene period 95, 98–9
  explosive growth of 163–4
  fossil evidence 99
  glaciation and 35
  human activity and 2, 166–7,
    187–90, *188*, 243
  ice core samples 163, 164
  ice melting after Last Glacial
    Maximum 146
  Keeling curve 161–4, *162*
  Phanerozoic fluctuations 50–1
  plate subduction 22–3
  predicting emissions 243
  tundra warming and 203–4
  volcanic emissions 20, 27, 51
  in water 21
  water vapour and 114
  weathering rocks 51, 59–60
  wildfires 183
Caricchi, Luca 121
Catalina–Jemez, New Mexico and
  Arizona 223
catastrophic events
  asteroid impacts 61–3
  'catastrophism' 48
  mass extinctions 60–3
  pre-Cambrian possibility 47
  super-volcanoes 60–1
cement production 164
Cenozoic period
  carbon dioxide 135

changes from Eocene 117–18
drop in temperature 117
'Himalayan cooling' 120–1
Mediterranean dries up 123–4
Cerling, Thure 122
Chamberlin, Thomas 51
Chaos: A Very Short Introduction
  (Smith) 234
Charney, Jule
  Eocene albedo 98
  green Sahara 151
  mechanism 151
  vegetation and climate 69–70
  weather prediction 242
Charpentier, Jean de 125, 126
chemistry *see* geochemistry
Chibbaro, Sergio
  Reductionism, Emergence and Levels
    of Reality (with Rondoni and
    Vulpiani) 234
China, Arctic sea ice and 196–7
Chinese Ecosystem Research Network
  (CERN) 224
chlorofluorocarbons, ozone layer and
  13–14
Cita, Bianca Maria 123
Clark, Peter 136–7, 146
Clausius (Rudolf) – Clapeyron (Émile)
  law 36–7
  clouds and greenhouse effect 113–14
  maximum water vapour 105
climate
  Cambrian fluctuations 50–2
  causes, effects and responses 134–8
  comparative climatology 86
  complexity of system 38
  continually changing 1
  critical thresholds 49
  'Critical Zone' ecosystem 72–3
  historical climatology 155–6
  Holocene 149–57
  influence of vegetation 69–70
  operational climatology 252–3